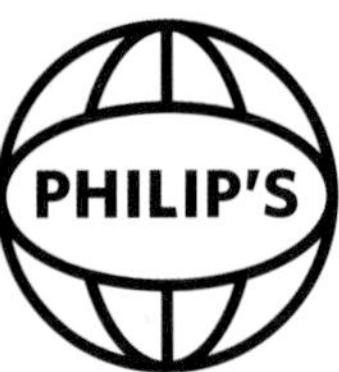

BRITAIN
& IRELAND

NIGEL
HENBEST

STARGAZING 2025

MONTH-BY-MONTH GUIDE TO THE NIGHT SKY

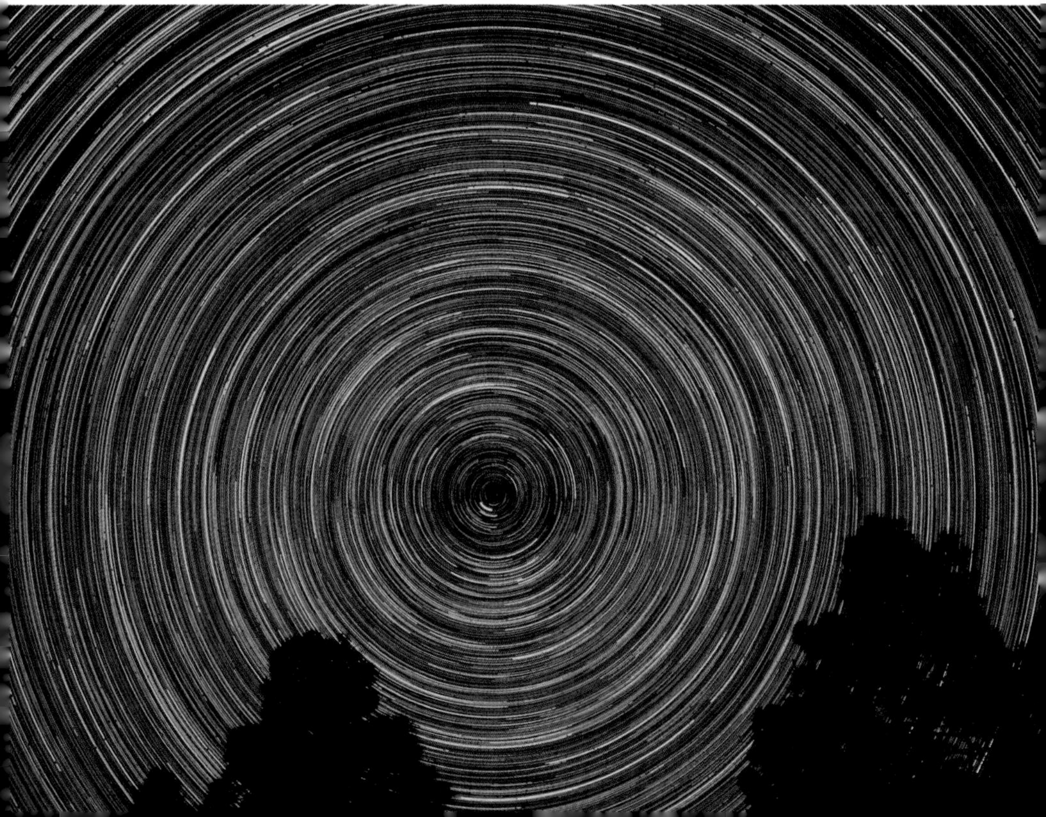

www.philips-maps.co.uk

Published in Great Britain in 2024 by Philip's,
a division of Octopus Publishing Group Limited
(www.octopusbooks.co.uk)
Carmelite House, 50 Victoria Embankment,
London EC4Y 0DZ
An Hachette UK Company (www.hachette.co.uk)

ISBN 978-1-84907-652-4

A CIP catalogue record for this book is available from the British Library.

Printed in China

Cover: The Milky Way at St Michael's Mount, Cornwall.
Title page: Star trails at Mirador Astronómico Llana del Jable, Tenerife, Canary Islands.

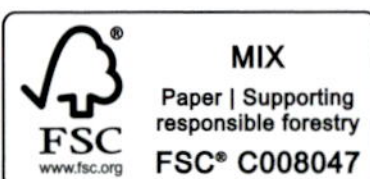

CONTENTS

Welcome to the latest edition of *Stargazing*! Within these pages, you'll find a complete guide to everything happening in the night sky throughout 2025 – whether you're a beginner or an experienced astronomer.

With the 12 monthly Star Charts, you can find your way around the sky on any night in the year. Impress your friends by identifying celestial sights ranging from the brightest planets to some pretty obscure constellations.

Every page of *Stargazing 2025* is bang up-to-date, bringing you everything that's new this year, from shooting stars to eclipses. And I'll start with a run-down of the most exciting sky sights on view in 2025 (opposite).

THE STAR CHARTS

A reliable map is just as essential for exploring the heavens as it is for visiting a foreign country. So each monthly section starts with a circular **Star Chart** showing the whole evening sky.

To keep the maps uncluttered, I've plotted about 200 of the brighter stars (down to third magnitude), which means you can pick out the main star patterns – the constellations. (If the charts showed every star visible on a really dark night, there'd be around 3000 stars on each!) I also show the ecliptic: the apparent path of the Sun in the sky; it's closely followed by the Moon and planets as well.

You can use these charts throughout the UK and Ireland, along with most of Europe, North America and northern Asia – between 40 and 60 degrees north – though the precise timings apply specifically to Britain and Ireland.

USING THE STAR CHARTS

It's pretty easy to use the charts. Start by working out your compass points. South is where the Sun is highest in the sky during the day; east is roughly where the Sun rises, and west where it sets. At night, you can find north by locating the Pole Star – Polaris – by using the stars of the Plough (see July's Constellation).

The left-hand chart then shows your view to the north. Most of the stars here are visible all year: these circumpolar constellations wheel around Polaris as the seasons progress. Your view to the south appears in the right-hand chart; it changes much more as the Earth orbits the Sun. Leo's prominent 'Sickle' is high in the spring skies. Summer is dominated by bright Vega, Deneb and Altair. Autumn's familiar marker is the Square of Pegasus; while the stars of Orion rule the winter sky.

During the night, our perspective on the sky also alters as the Earth spins round, making the stars and planets appear to rise in the east and set in the west. The charts depict the sky in the late evening (the exact times are noted in the captions). As a rule of thumb, if you are observing two hours later, then the following month's chart will be a better guide to the stars on view – though beware: the Moon and planets won't be in the right place.

THE PLANETS, MOON AND SPECIAL EVENTS

The charts also highlight the **planets** above the horizon in the late evening. I've

HIGHLIGHTS OF THE YEAR

- **3 January:** Venus is close to the Moon.
- **Night of 3/4 January:** excellent views of the Quadrantid meteor shower.
- **4 January:** the Moon occults Saturn.
- **10 January, early hours:** the Moon occults the Pleiades.
- **12 January:** Mars is nearest to the Earth.
- **Night of 13/14 January:** the Full Moon passes only 12 arcminutes from Mars.
- **16 January:** Mars is opposite the Sun.
- **1 February:** the crescent Moon near Venus, Neptune and Saturn.
- **9 February:** the Moon skims just 7 arcminutes from Mars.
- **19 February:** Venus at its brightest in the evening sky.
- **24 and 25 February:** Mercury passes Saturn.
- **2 March:** Venus forms a stunning sight with the crescent Moon.
- **Night of 8/9 March:** the Moon passes only 40 arcminutes from Mars.
- **14 March:** a total eclipse of the Moon is visible from the UK and Ireland as the Moon sets.
- **29 March:** a partial eclipse of the Sun is visible across Ireland and the UK.
- **Night of 1/2 April:** the Moon occults the Pleiades.
- **22 April, before dawn:** Venus at its brightest in the morning sky.
- **Night of 22/23 April:** it's a great year for observing the Lyrid meteor shower.
- **3 May:** the Moon passes Mars and Praesepe.
- **3–5 May:** Mars moves through Praesepe.
- **28 May:** Jupiter lies below the Moon.
- **Night of 10/11 June:** the lowest Full Moon in 19 years.
- **17 June:** Mars 45 arcminutes from Regulus.
- **22 June, before dawn:** Venus lies near the crescent Moon.
- **23 June, early hours:** the crescent Moon occults the Pleiades.
- **21–23 July, before dawn:** the crescent Moon passes Venus and Jupiter.
- **12 August, before dawn:** Venus passes only 52 arcminutes from Jupiter.
- **20 August, before dawn:** the crescent Moon lies between Venus and Jupiter.
- **7 September:** a total lunar eclipse is visible from the UK and Ireland as the Moon rises.
- **12 September:** the Moon occults the Pleiades.
- **16–17 September, early hours:** the Moon lies near Jupiter, Castor and Pollux.
- **19 September, before dawn:** Venus is only 48 arcminutes from Regulus, near the Moon.
- **21 September:** Saturn is opposite the Sun and nearest to the Earth.
- **22–23 September:** Neptune is nearest to the Earth and opposite the Sun.
- **8 October, before dawn:** the Draconid meteor shower may produce bright fireballs.
- **10 October, before dawn:** the Moon occults the Pleiades.
- **Night of 21/22 October:** an excellent view of the Orionid meteor shower.
- **5 November:** the biggest and brightest supermoon this year; the best since 2019.
- **Night of 17/18 November:** a great year for observing the Leonid meteor shower.
- **21 November:** Uranus is at its closest to the Earth and opposite the Sun.
- **Night of 3/4 December:** the Moon occults the Pleiades.
- **7 December:** the Moon is near Jupiter, Castor and Pollux.
- **10 December, before dawn:** the Moon occults Regulus.
- **Night of 13/14 December:** an excellent year for observing the Geminid meteor shower.

indicated the track of any **comets** known at the time of writing; though I can't guide you to a comet that's found after the book has been printed!

I've plotted the position of the Full Moon each month, and also the **Moon's position** at three-day intervals before and afterwards. If there's a **meteor shower** in the month, the charts show its radiant – the position from which the meteors stream outwards.

The **Calendar** provides a daily guide to the Moon's phases and other celestial happenings. I've detailed the most interesting in the **Special Events** section, including close pairings of the planets, times of the equinoxes and solstices and – most exciting – **eclipses** of the Moon and Sun.

Check out the **Planet Watch** page for more about the other worlds of the Solar System, including their antics at times they're not on the monthly Star Charts. I've illustrated unusual planetary and lunar goings-on in the **Planet Event Charts**. There's a full guide to planetary motions, eclipses and meteor showers in **Solar System 2025** on pages 80–82.

MONTHLY OBJECTS, TOPICS AND PICTURES

Each monthly section highlights a fascinating **object** – a planet, a star or a nebula – as well as a stunning **picture** taken by an amateur based in Britain or Ireland along with full technical information. There's also an in-depth exploration of an intriguing and often newsworthy **topic**, ranging from collisions between giant galaxies to the growing 'constellations' of satellites that ruin our view of the night sky.

GETTING IN DEEP

A practical **Observing Tip** is included each month, helping you to explore the sky with the naked eye, binoculars or a telescope.

Check out my guide to the **Top 20 Sky Sights**, including nebulae, star clusters and galaxies. You'll find it on pages 83–85.

And equipment expert Robin Scagell reviews a new breed of telescope that's sweeping the amateur astronomy world. Small refractors resemble the traditional lens-telescope, but they feature state-of-the-art optics that allow you to take superlative images of nebulae and galaxies.

The final section details the internationally approved dark-sky sites in Britain and Ireland, where you're guaranteed to be free of light pollution. It also includes a round-up of planetariums and public observatories you can visit, plus a guide to the best star parties and astronomical festivals.

Happy stargazing!

JARGON BUSTER

Have you ever wondered how astronomers describe the brightness of the stars or how far apart they appear in the sky? Not to mention how we can measure the distances to the stars? If so, you can quickly find yourself mired in some arcane astro-speak – magnitudes, arcminutes, light years and the like.

Here's a quick and easy guide to busting the astronomical jargon.

Magnitudes

It only takes a glance at the sky to see that some stars are pretty brilliant, while many more are dim. But how do we describe to other people how bright a star appears?

Around 2000 years ago, ancient Greek astronomers ranked the stars into six classes, or **magnitudes**, depending on their brightness. The most brilliant stars were first magnitude, and the faintest stars you can see came in at sixth magnitude. So the stars of the Plough, for instance, are second magnitude while the individual Seven Sisters in the Pleiades are fourth magnitude.

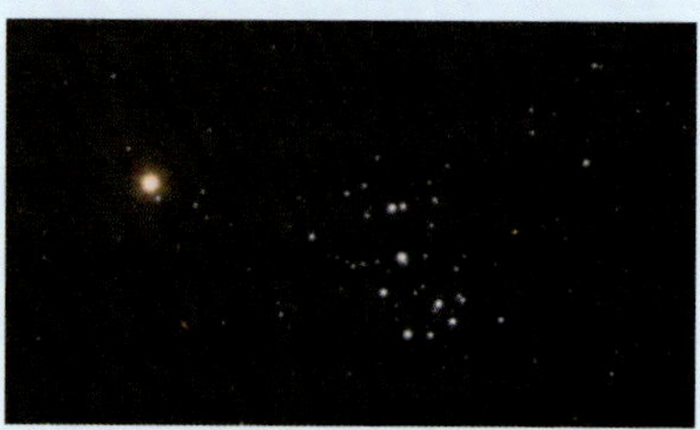

Mars (magnitude -1.6 here) shines a hundred times brighter than the Seven Sisters in the Pleiades, which are around 5 magnitudes fainter.

Today, scientists can measure the light from the stars with amazing accuracy. (Mathematically speaking, a difference of five magnitudes represents a difference in brightness of one hundred times.) So the Pole Star is magnitude +2.0, while Rigel is magnitude +0.1. Because we've inherited the ancient ranking system, the brightest stars have the *smallest* magnitude. In fact, the most brilliant stars come in with a negative magnitude, including Sirius (magnitude −1.5).

And we can use the magnitude system to describe the brightness of other objects in the sky, such as stunning Venus, which can be almost as brilliant as magnitude −5. The Full Moon and the Sun have whopping negative magnitudes!

At the other end of the scale, stars, nebulae and galaxies with a magnitude fainter than +6.5 are too dim to be seen by the naked eye. Using ever larger telescopes – or by observing from above Earth's atmosphere – you can perceive fainter and fainter objects. The most distant galaxies visible to the Hubble Space Telescope are ten billion times fainter than the naked-eye limit.

Here's a guide to the magnitude of some interesting objects:

Sun	−26.7
Full Moon	−12.5
Venus (at its brightest)	−4.9
Sirius	−1.5
Betelgeuse (variable)	0.0 – +1.6
Polaris (Pole Star)	+2.0
Faintest star visible to the naked eye	+6.5
Faintest star visible to the Hubble Space Telescope	+31

Degrees of separation

Astronomers measure the distance between objects in the sky in **degrees** (symbol °): all around the horizon is 360°, while it's 90° from the horizon to the point directly overhead (the zenith).

As you can see in the photograph, it's possible to use your hand – held at arm's length – to give a rough idea of angular distances in the sky.

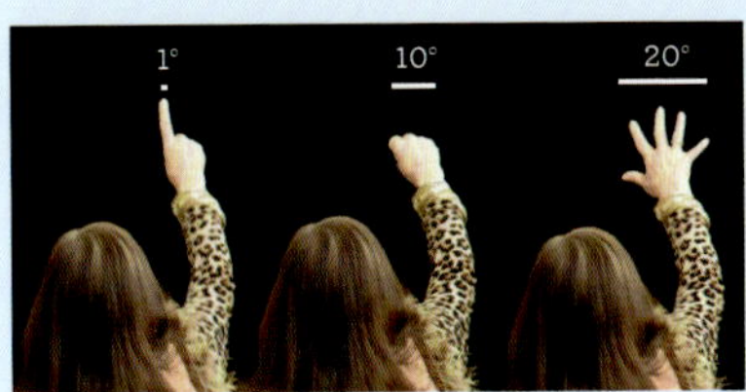

For objects that are very close together – like many double stars – we divide the degree into 60 arcminutes (symbol '). And for celestial objects that are extremely tiny – such as the discs of the planets – we split each arcminute into 60 arcseconds (symbol "). To give you an idea of how small these units are, it takes 3600 arcseconds to make up one degree.

Here are some typical separations and sizes in the sky:

Length of the Plough	25°
Width of Orion's Belt	3°
Diameter of the Moon	31'
Separation of Mizar and Alcor	12'
Diameter of Jupiter	45"
Separation of Albireo A and B	35"

How far's that star?

Everything we see in the heavens lies a long way off. We can give distances to the planets in millions of kilometres. But the stars are so distant that even the nearest, Proxima Centauri, lies some 40 million million kilometres away. To turn those distances into something more manageable, astronomers use a larger unit: one **light year** is the distance that light travels in a year.

One light year is about 9.46 million million kilometres. That makes Proxima Centauri a much more manageable 4.2 light years away from us. Here are the distances to some other familiar astronomical objects, in light years:

Sirius	8.6
Polaris	440
Centre of the Milky Way	26,700
Andromeda Galaxy	2.5 million
Most distant galaxies seen by James Webb Space Telescope	13.5 billion

- The sky at 10 pm in mid-January, with Moon positions at three-day intervals either side of Full Moon.
- The star positions are also correct for 11 pm at the beginning of January, and 9 pm at the end of the month.
- The planets move slightly relative to the stars during the month.

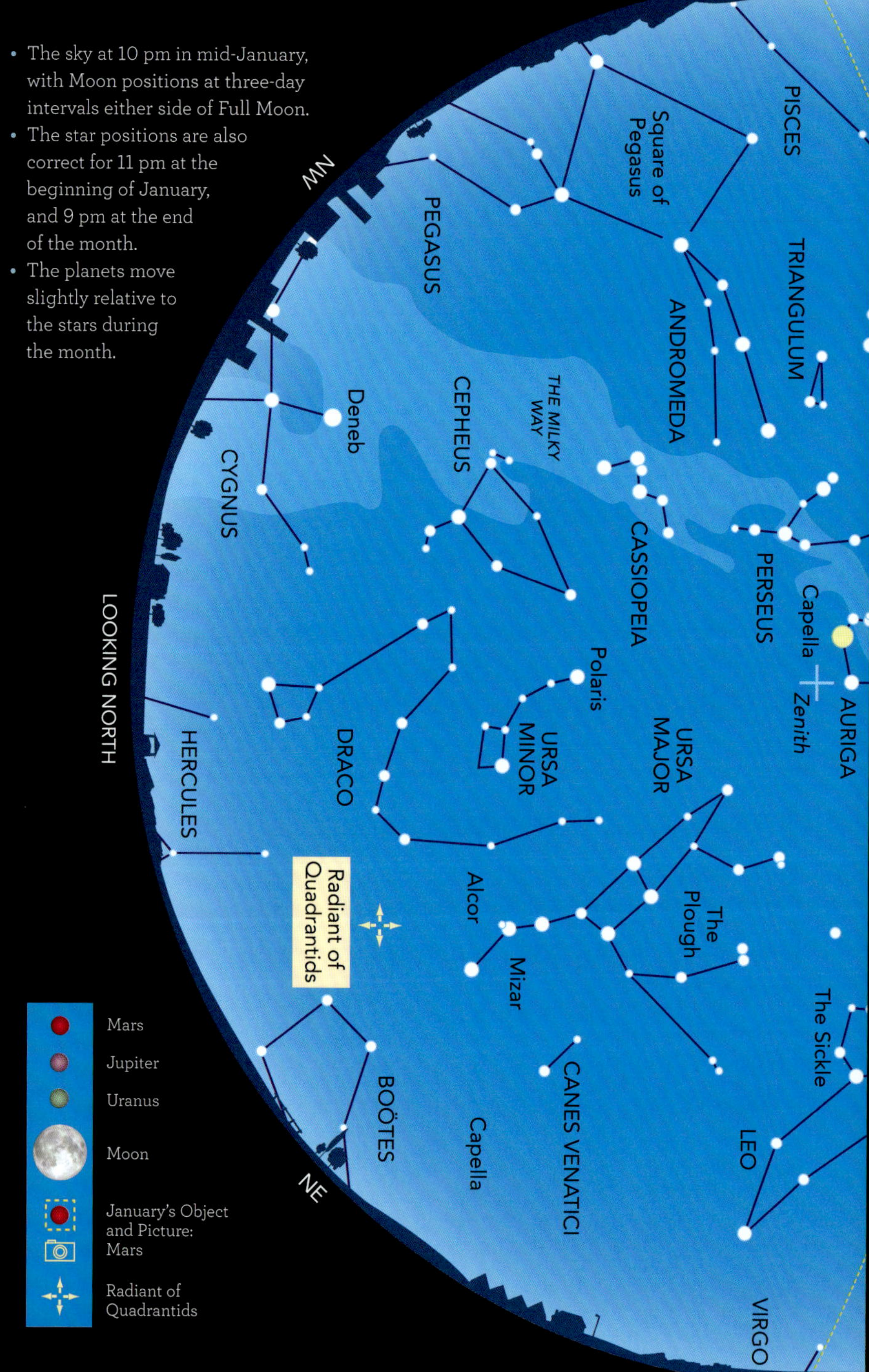

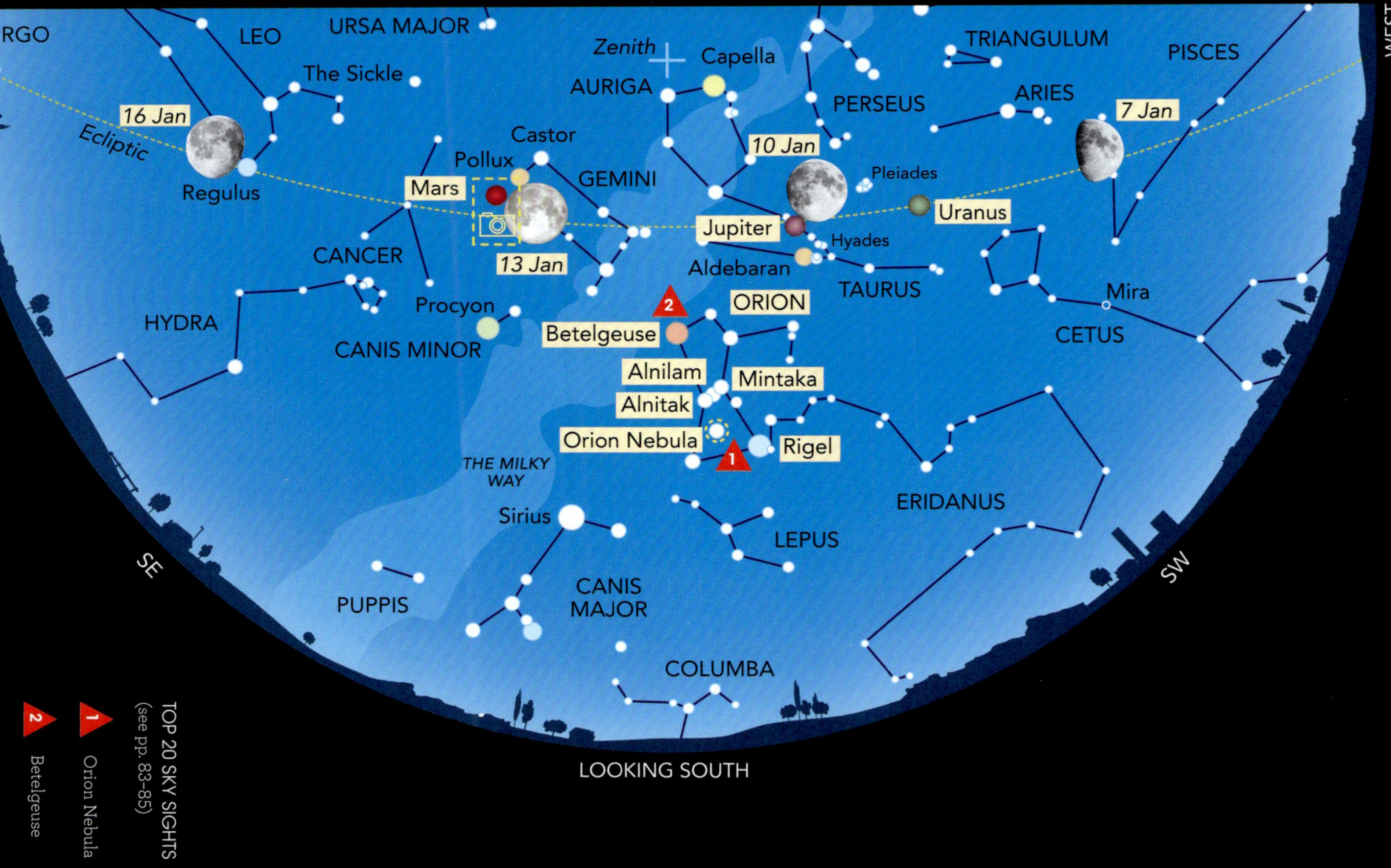

TOP 20 SKY SIGHTS
(see pp. 83–85)

1 Orion Nebula

2 Betelgeuse

JANUARY

It's an action-packed start to 2025! The three most brilliant planets are all on view in the evening sky, with **Mars** at its nearest for two years. Add to that the Moon moving in front of Saturn and then the Seven Sisters, and also a spectacular display of shooting stars, and it's going to be a month to remember.

JANUARY'S CONSTELLATION

Opening the new year with a flourish, the scintillating stellar figure of **Orion** strides high across the heavens this month. According to various ancient cultures, this constellation represented a shepherd, a mighty king or even a canoe but to the ancient Greeks he was a formidable hunter, and also the world's most handsome man. In the sky Orion also cuts a dash, containing one-tenth of the brightest stars in the sky.

Blood-red **Betelgeuse** marks one of his shoulders. This red giant is 760 times wider than the Sun, and one day it's destined to blow itself apart as a supernova. Orion's other brilliant star, **Rigel**, lies at the bottom of the hunter's tunic. In contrast to **Betelgeuse**, it's blue-white in colour, with a searingly hot temperature of 12,000°C.

OBSERVING TIP

The brilliant winter constellations and planets are tempting us to spend plenty of time outside. So make sure you dress up warmly! Lots of layers are better than just a heavy coat, as they trap more air close to your skin. Heavy-soled boots with two pairs of socks stop the frost creeping up your legs. And a surprising amount of your body heat escapes through the top of your head, so complete your outfit with a beanie or – even better – a hat with ear flaps.

Orion's Belt is marked by well-matched **Alnitak**, **Alnilam** and **Mintaka**, lying over 1200 light years away. Despite this immense distance, they're prominent in our skies because each outshines our Sun 200,000 times over.

Look carefully below the Belt to spot a faint patch of light representing Orion's sword. This is the great **Orion Nebula**, a seething maelstrom of incandescent gas that's the birthplace of new stars created from a dark cloud of dust and gas.

JANUARY'S OBJECT

Mars is the only planet on which we can clearly discern surface details from the Earth (see Picture). Like our world, the rocky Red Planet has polar caps, seasons, an atmosphere (albeit very thin) and even clouds.

For a planet only half the Earth's size, its geology is truly astonishing. Mars boasts an enormous canyon, Valles Marineris, that is 4000 kilometres long and 7000 metres deep – big enough to swallow the Alps, with room to spare!

With an altitude of 26,000 metres, Mars's giant dormant volcano Olympus Mons is three times the height of Mount Everest. It would completely cover England, with a central crater that could contain London several times over.

And there's abundant evidence that the now-dry planet was once awash with rivers, lakes and maybe even oceans. Water is the cradle where life begins,

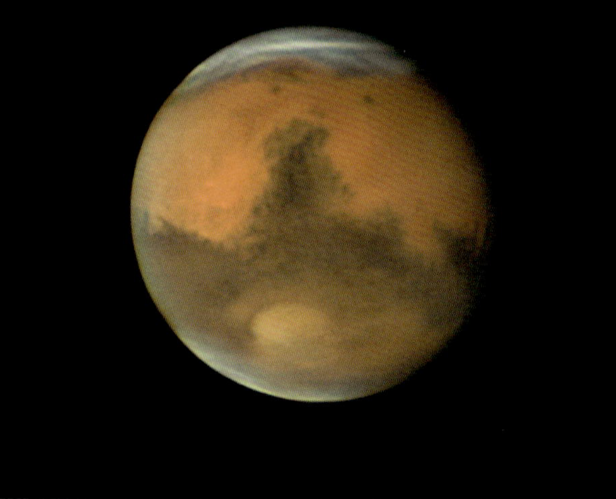

During the period when Mars was last close to the Earth, Damian Peach imaged the Red Planet on 21 November 2022 with a Player One Uranus-C camera on a Celestron C14 355-mm Schmidt-Cassegrain telescope.

and the elixir that sustains all living things. Scientists hope that samples now being collected by the Perseverance Mars rover will reveal whether life ever flourished on the Red Planet.

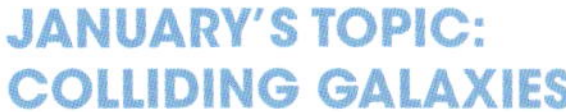

JANUARY'S TOPIC: COLLIDING GALAXIES

The Universe is filled with galaxies, giant star cities like our Milky Way. They look serene and stately, but galaxies are typically speeding through space at around a million kilometres per hour. If another galaxy lies in the way, the outcome will be messy...

In the case of M51, it has been a glancing blow. A smaller galaxy whizzed through the edge of its wide disc, the energy of the collision stirring up the larger galaxy's material into a prominent shape that has led to M51's nickname, the Whirlpool Galaxy.

The Antennae Galaxy is the outcome of a head-on smash. The impact has ejected stars into a pair of long streamers that inspire its nickname. As gas clouds in the centres of the two galaxies have collided, the shock waves have spawned a massive outbreak of starbirth, their energy adding to the cosmic mayhem.

With the spare gas in a galaxy collision compressed into stars, the end result of a cosmic collision is a larger galaxy that is devoid of interstellar material: the kind of boring galaxy that astronomers call an elliptical. Astronomers now think that most of the elliptical galaxies in the Universe are the wreckage of cosmic collisions.

JANUARY'S PICTURE

It looks at first sight like an image from the Hubble Space Telescope, or even a visiting spacecraft, but this exquisitely detailed view of **Mars** was obtained from a back garden in Selsey, Hampshire!

Central in Damian Peach's image is the Red Planet's most prominent feature, Syrtis Major, formed from dark lava that has oozed out from ancient volcanic vents. The pale oval below is Hellas, the biggest impact crater on Mars. The bottom of Hellas is the lowest point on Mars, with an atmospheric pressure high enough that it's the only place where water could exist as a liquid.

There's haze over both the poles, and at the top you can see the gleaming white north polar cap. While the arctic regions of the Earth are covered in ice, Mars is so cold that its polar caps are made of frozen carbon dioxide.

JANUARY'S CALENDAR

SUNDAY	MONDAY	TUESDAY	WEDNESDAY	THURSDAY	FRIDAY	SATURDAY
			1	2	3 Moon near Venus; Quadrantids	4 Quadrantids (am); Earth at perihelion; Moon occults Saturn
5	6 11.56 pm First Quarter Moon	7	8	9	10 Moon occults Pleiades (am); Venus E elongation; Moon near Jupiter	11
12 Mars closest to Earth	13 10.27 pm Full Moon near Mars	14 Moon very near Mars (am)	15	16 Mars opposition; Moon near Regulus	17	18 Venus near Saturn
19	20	21 Moon near Spica (am); 8.31 pm Last Quarter Moon	22	23	24	25 Moon near Antares (am)
26	27	28	29 12.36 pm New Moon	30	31	

SPECIAL EVENTS

- **3 January:** there's a beautiful pairing of Venus and the crescent Moon as night falls (Chart 1a).
- **Night of 3/4 January:** the maximum of the **Quadrantid meteor shower**, dust particles from the old comet 2003 EH_1 burning up in the Earth's atmosphere. It's an excellent year for catching these bright colourful shooting stars after the Moon has set at 8.30 pm.
- **4 January, 1.28 pm:** the Earth is at perihelion, its closest point to the Sun (147 million km away).
- **4 January, early evening:** the Moon moves in front of Saturn (Chart 1b). The occultation starts between 5.13 and 5.23 pm (the exact time depending on your location) when the dark edge of the Moon hides Saturn. The planet emerges from the Moon's bright side between 6.20 and 6.31 pm.
- **10 January, 1.30–4.00 am:** the Moon slices across the lower part of the Pleiades, hiding four of the bright Seven Sisters – Merope, Alcyone, Atlas and Pleione – and many of the fainter stars.
- **10 January:** Venus is at its greatest separation from the Sun. The bright 'star' near the Moon is Jupiter.
- **12 January:** Mars is nearest to the Earth at 96 million km (see Planet Watch).
- **Night of 13/14 January:** the Full Moon passes only 12 arcminutes from Mars, with Castor and Pollux nearby.
- **16 January:** Mars is opposite to the Sun (see Planet Watch).

JANUARY'S PLANET WATCH

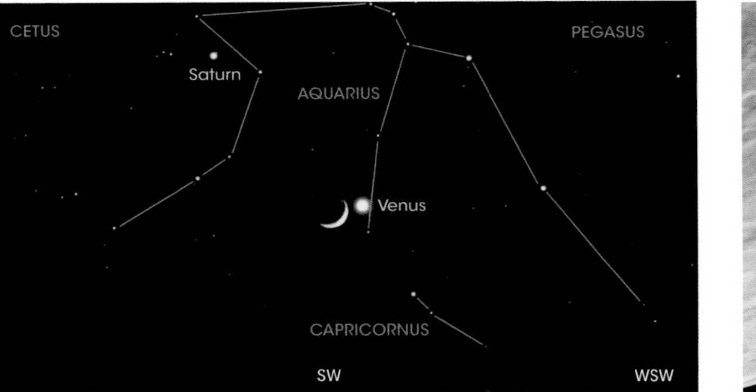

1a *3 January, 6.30 pm. Venus and the crescent Moon, with Saturn.*

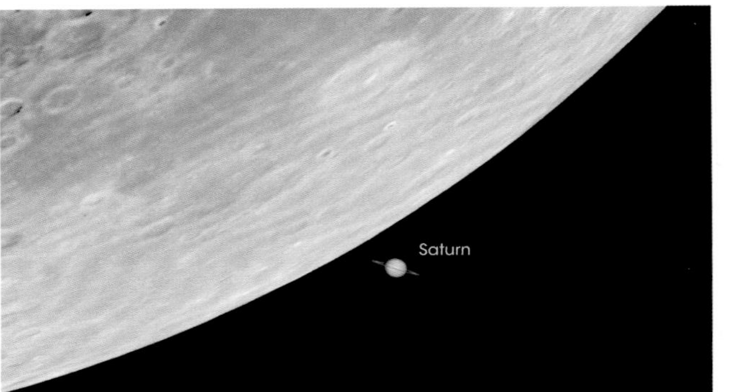

1b *4 January, 6.30 pm. Saturn emerges from lunar occultation.*

- **Venus** is a brilliant Evening Star in the south-west after sunset: at magnitude –4.6, it's visible long before any stars can be seen, and it sets around 8.45 pm. The crescent Moon passes just to the left of Venus on 3 January (Chart 1a). The planet reaches its greatest separation from the Sun on 10 January. Through a small telescope, you can see Venus as a half-lit globe.
- At the start of January, **Saturn** lies to the upper left of Venus, in Aquarius. At magnitude +1.1, it's 200 times fainter than the Evening Star. The Moon occults Saturn on 4 January (see Special Events and Chart 1b). As Venus moves upwards, it passes to the right of Saturn on 18 January. By the end of the month Saturn sets at 8 pm.
- Second only to Venus in brightness, **Jupiter** (magnitude –2.6) lies near Aldebaran in Taurus, and sets around 5 am. You'll find the Moon near the giant planet on 10 January.
- Though only the third-brightest planet on view, at magnitude –1.5 **Mars** is still more resplendent than any of the stars. On 12 January, the Red Planet is nearer to the Earth than it's been since 2022, and it's directly in line with the Sun and the Earth a few days later, on 16 January. The Full Moon skims past Mars in the early hours of 14 January. Visible all night long, the Red Planet moves from Cancer to Gemini mid-month; it's in line with Castor and Pollux on 17 January.
- **Neptune** (magnitude +7.8) lies in Pisces and sets around 10 pm, while **Uranus** – in Aries – sets about 3.30 am and glows at magnitude +5.7.
- During the first week of January, you may catch **Mercury** before dawn, very low in the south-east. At magnitude –0.4, the planet rises around 6.40 am.

- The sky at 10 pm in mid-February, with Moon positions at three-day intervals either side of Full Moon.
- The star positions are also correct for 11 pm at the beginning of February, and 9 pm at the end of the month.
- The planets move slightly relative to the stars during the month.

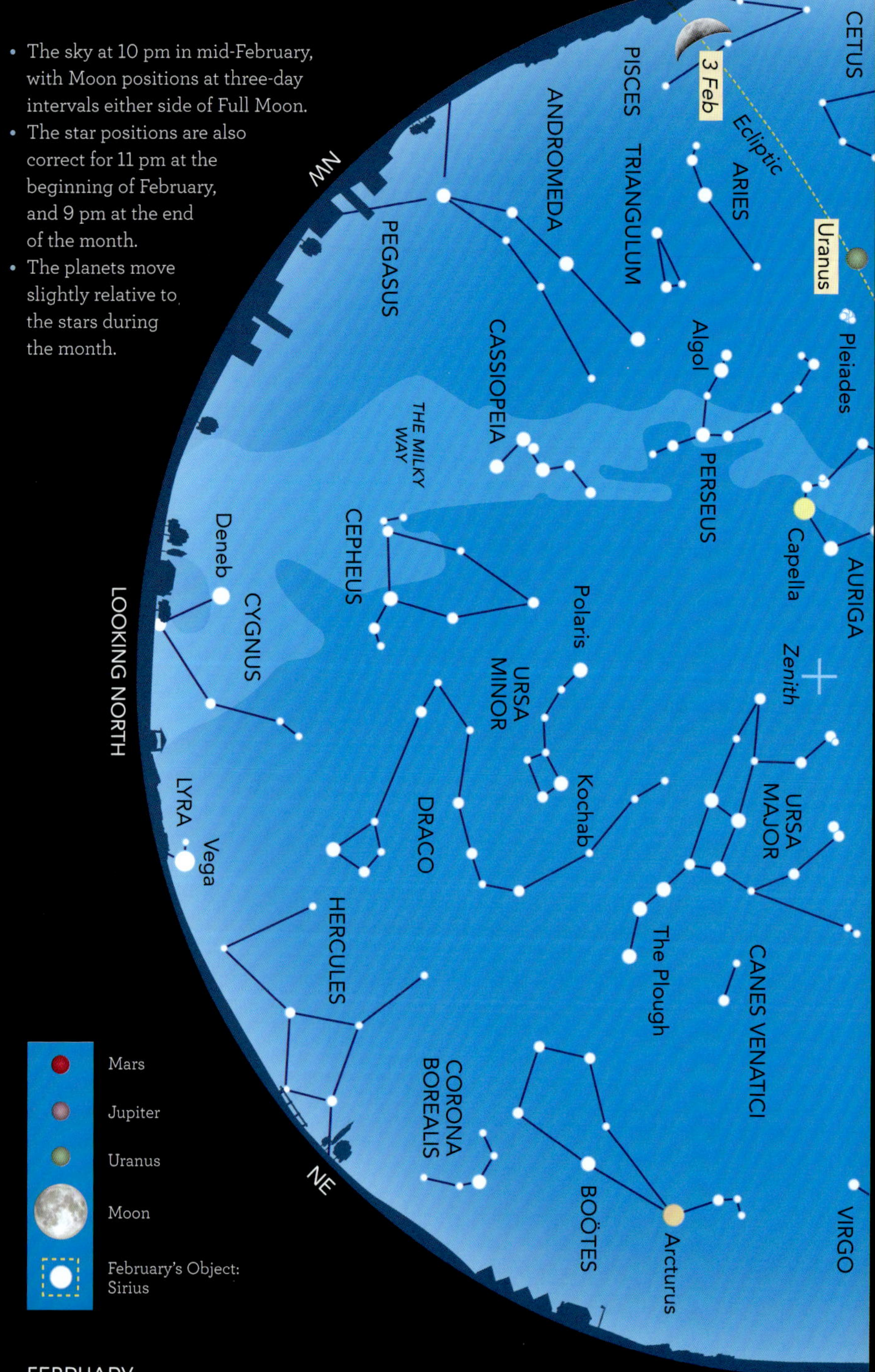

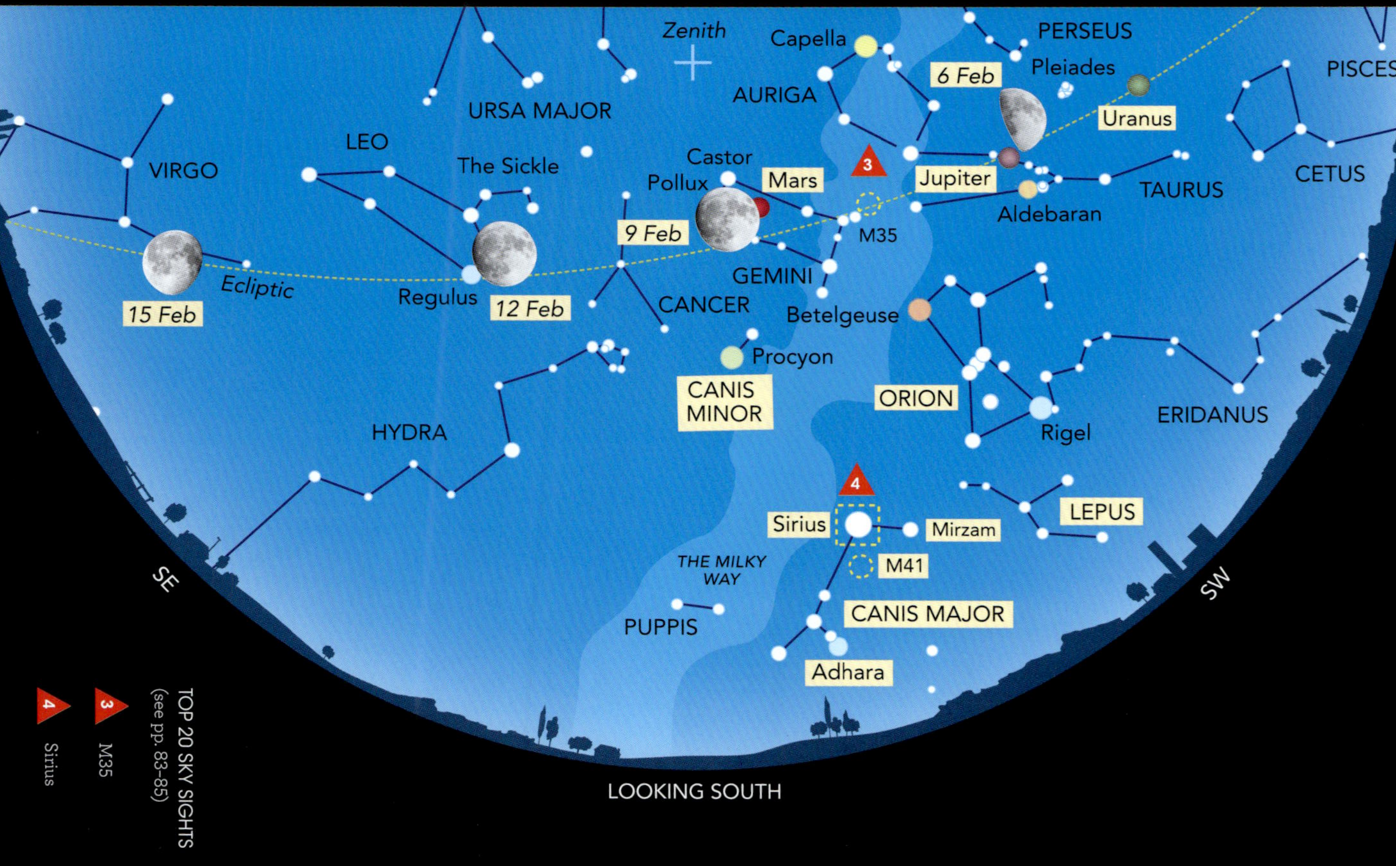

FEBRUARY

TOP 20 SKY SIGHTS
(see pp. 83–85)

3 M35

4 Sirius

This month, catch all the planets of the Solar System on display during the evening. Venus, **Mars** and **Jupiter** are blazing beacons that you can't miss. Saturn and Mercury are visible to the unaided eye, low in the evening twilight. With binoculars or a small telescope, seek out faint **Uranus** and Neptune. For the final planet, just look down between your feet!

FEBRUARY'S CONSTELLATION

Canis Major is the larger of **Orion**'s two hunting dogs – the other, naturally, being **Canis Minor** – and it's crowned by the Dog Star, brilliant **Sirius** (see Object). The Great Dog is chasing faint **Lepus** (the Hare), cowering below Orion.

To the right of Sirius lies **Mirzam**, whose Arabic name means 'the announcer' as it rises just before Sirius. Prominent **Adhara** (magnitude +1.5) only just fails to make the list of first-magnitude stars. Five million years ago, Adhara was the brightest star in the sky as it passed close to the Sun, blazing as brightly as Venus.

The beautiful star cluster **M41** is easily visible through binoculars, and even to the unaided eye. The Greek philosopher Aristotle, in 325 BC, referred to 'a cloudy spot' in Canis Major: if that was M41 it's the earliest surviving description of a deep-sky object.

OBSERVING TIP

Venus is a real treat this month. And – if you can get to view it through a small telescope – you'll see it as a small glowing crescent. But don't wait for the sky to get totally dark, because the cloud-wreathed world will appear so brilliant it's difficult to make out any details. You're best off viewing Venus when the Evening Star first becomes visible in the twilight glow. Through a telescope, the planet then appears less dazzling against a pale blue sky.

FEBRUARY'S OBJECT

Reigning supreme over all the stars in our sky is dazzling **Sirius**, at the head of Canis Major (see Constellation), resplendent at magnitude –1.5. The Dog Star isn't particularly luminous, though, shining only 25 times more brightly than the Sun. It just happens to lie nearby on the cosmic scale, a mere 8.6 light years away.

The ancient Greeks named this star Seirios ('scorcher'), and they believed its rays added to the Sun's heat in summer, to create the hot humid 'dog days' when people were lethargic and dogs went crazy. To the Egyptians, the appearance of Sirius at dawn heralded the Nile floods, a welcome harbinger for a bumper harvest.

Boasting a temperature of almost 10,000°C, Sirius is twice as heavy as the Sun. And it's relatively young: just 230 million years old, as compared to the Sun's venerable 4600 million years. Sirius has a faint white dwarf companion, the Pup (see December's Topic).

FEBRUARY'S TOPIC: FRITZ ZWICKY (1898–1974)

Swiss astronomer Fritz Zwicky is one of the great unsung scientific heroes. Though little known except to astronomers, he pioneered many of the hottest topics in the Cosmos today.

After graduating from the Federal Institute of Technology in Zurich, Zwicky crossed the Atlantic to use the great telescopes in California. In 1933, he discovered that galaxies in the Coma galaxy cluster are speeding so fast that they should all have escaped by now. Zwicky deduced that the gravity of something invisible must be reining them in. He named it 'dark matter'; and we now know that dark matter constitutes most of the material in the Universe.

With his German colleague Walter Baade (1893–1960), Zwicky was the first to realise that some erupting stars were far more powerful than traditional novae, and instead marked the total destruction of those stars. He called these stellar explosions 'supernovae'.

And a particular kind of supernova that Zwicky discovered in 1937 – Type Ia – was the key to astronomers in the 1990s deducing that a strange force – 'dark energy' – is ripping our Universe apart.

Zwicky's obscurity is partly due to his dislike of his American colleagues; he referred to them as 'spherical bastards', because they were bastards whatever way you viewed them! But his friends remember Zwicky for his generosity, for instance when he helped to rehome orphans after World War II.

FEBRUARY'S PICTURE

The shimmering iridescent curtains of an aurora are always spectacular, but the awe was notched up to another level when photographer Ian Sproat caught this display as a backdrop to the iconic 150-year-old tree in Sycamore Gap, on Hadrian's Wall in the north of England.

These multicoloured lights lie within the Earth's atmosphere, but the cause is astronomical. Storms on the Sun hurl electrically charged particles outwards through the Solar System. If they hit the Earth, our planet's magnetic field channels them towards the poles, where they light up our atmosphere like gas in a neon tube.

Sad to say, no-one will be taking a picture like this again. A few days after Ian's visit, the sycamore was maliciously cut down. 'It's like I was meant to see this,' Ian muses. With his companions, 'we gasped in awe of the beautiful lights above the tree – for one last time.'

'Sycamore Gap under the lights – the last show' was immortalised by Ian Sproat on 13 September 2023, with the help of a Sony Alpha 7 IV camera and a Sigma 20-mm f/1.4 DG DN lens. It was a 13-second exposure at f/2.2 and ISO 4000, and the noise was cleaned up in DxO software.

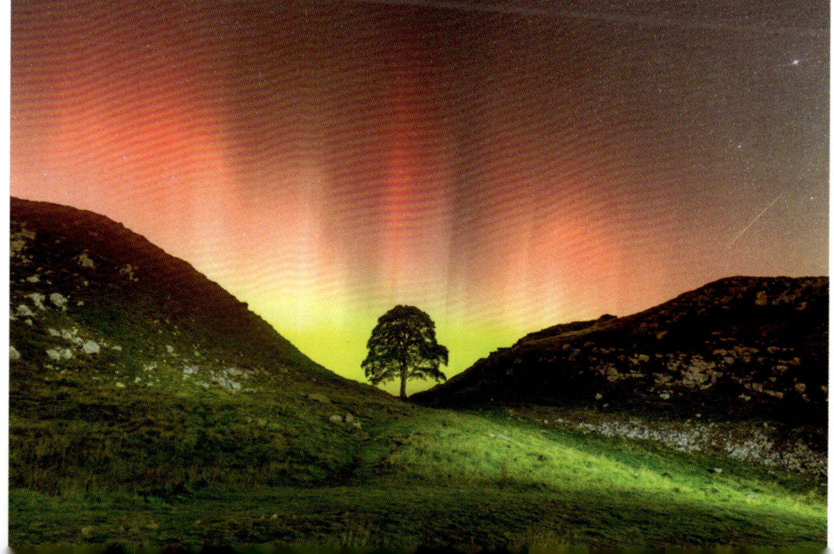

FEBRUARY'S CALENDAR

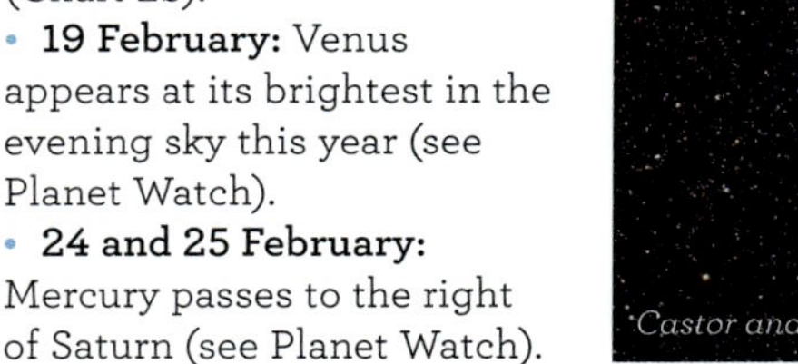

SUNDAY	MONDAY	TUESDAY	WEDNESDAY	THURSDAY	FRIDAY	SATURDAY
						1 Moon near Venus and Neptune
2	3	4	5 8.02 am First Quarter Moon	6 Moon near Jupiter	7	8
9 Moon very near Mars	10	11	12 1.53 pm Full Moon near Regulus	13	14	15
16	17	18	19 Venus at greatest brightness	20 5.32 pm Last Quarter Moon	21 Moon near Antares (am)	22
23	24 Mercury near Saturn	25 Mercury near Saturn	26	27	28 0.45 am New Moon	

SPECIAL EVENTS

- **1 February:** the narrow crescent Moon lies close to Venus in the evening sky, with Neptune nearby and Saturn below (see Planet Watch and Chart 2a).
- **6 February:** you'll find Jupiter to the left of the Moon.
- **9 February, 7 pm:** the Moon skims past Mars at a distance of only 7 arcminutes, with Castor and Pollux nearby (Chart 2b).
- **19 February:** Venus appears at its brightest in the evening sky this year (see Planet Watch).
- **24 and 25 February:** Mercury passes to the right of Saturn (see Planet Watch).

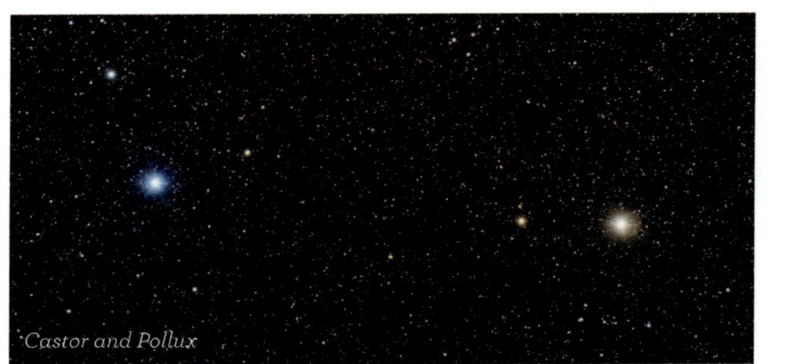

Castor and Pollux

FEBRUARY'S PLANET WATCH

2a 1 February, 6.30 pm. The Moon with Venus, Saturn and Neptune (brightness exaggerated).

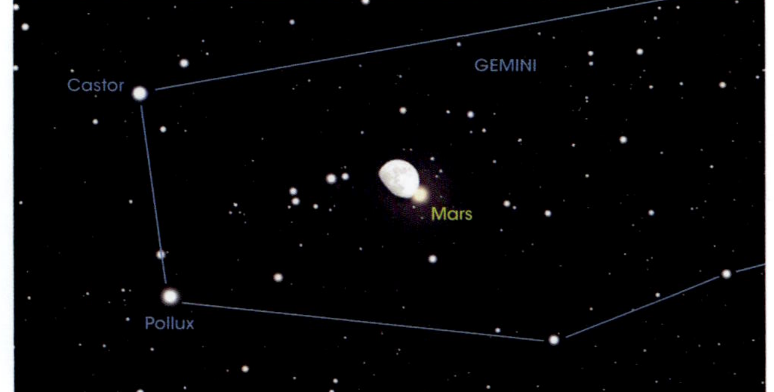

2b 9 February, 7 pm. The Moon skims past Mars, near Castor and Pollux.

- Reaching maximum brilliance (magnitude –4.9) on 19 February, **Venus** dazzles in the evening sky this month.
- The Evening Star sets at 9 pm, and forms a lovely sight with the crescent Moon on 1 February (Chart 2a). Through a telescope, Venus appears as a narrowing crescent.
- At the beginning of February, **Saturn** (magnitude +1.1) lies below Venus in Aquarius, setting about 8 pm. But by the month's end, it's disappeared into the twilight glow. Grab a telescope to view its famous rings almost edge-on to us: we're currently seeing them from the northern side, but when Saturn reappears from behind the Sun in May, we'll be viewing the rings from underneath.
- **Mercury** appears low in the evening sky towards the end of the month, setting about 7 pm. On 24 and 25 February, Mercury (magnitude –1.2) lies just to the right of fainter Saturn.
- **Neptune** lies in Pisces and sets about 8 pm. At magnitude +7.8, you'll need binoculars or a telescope to see it, but there's an easy chance to find it on 1 February, when Neptune forms the corner of a triangle with Venus and the Moon (Chart 2a).
- Its near-twin, **Uranus** is at the border of Aries and Taurus, at a dim magnitude +5.8. It sets around 1.30 am.
- Giant planet **Jupiter** (magnitude –2.4) lies in Taurus, near to Aldebaran, and sets about 3 am. The Moon passes nearby on 6 February.
- A ruby jewel adorning Gemini, **Mars** sets around 4 am. The Red Planet fades during February from magnitude –1.1 to magnitude –0.3. The Moon is very near Mars on 9 February (see Special Events).

- The sky at 10 pm in mid-March, with Moon positions at three-day intervals either side of Full Moon.
- The star positions are also correct for 11 pm at the beginning of March, and 10 pm at the end of the month (after BST and IST begin).
- The planets move slightly relative to the stars during the month.

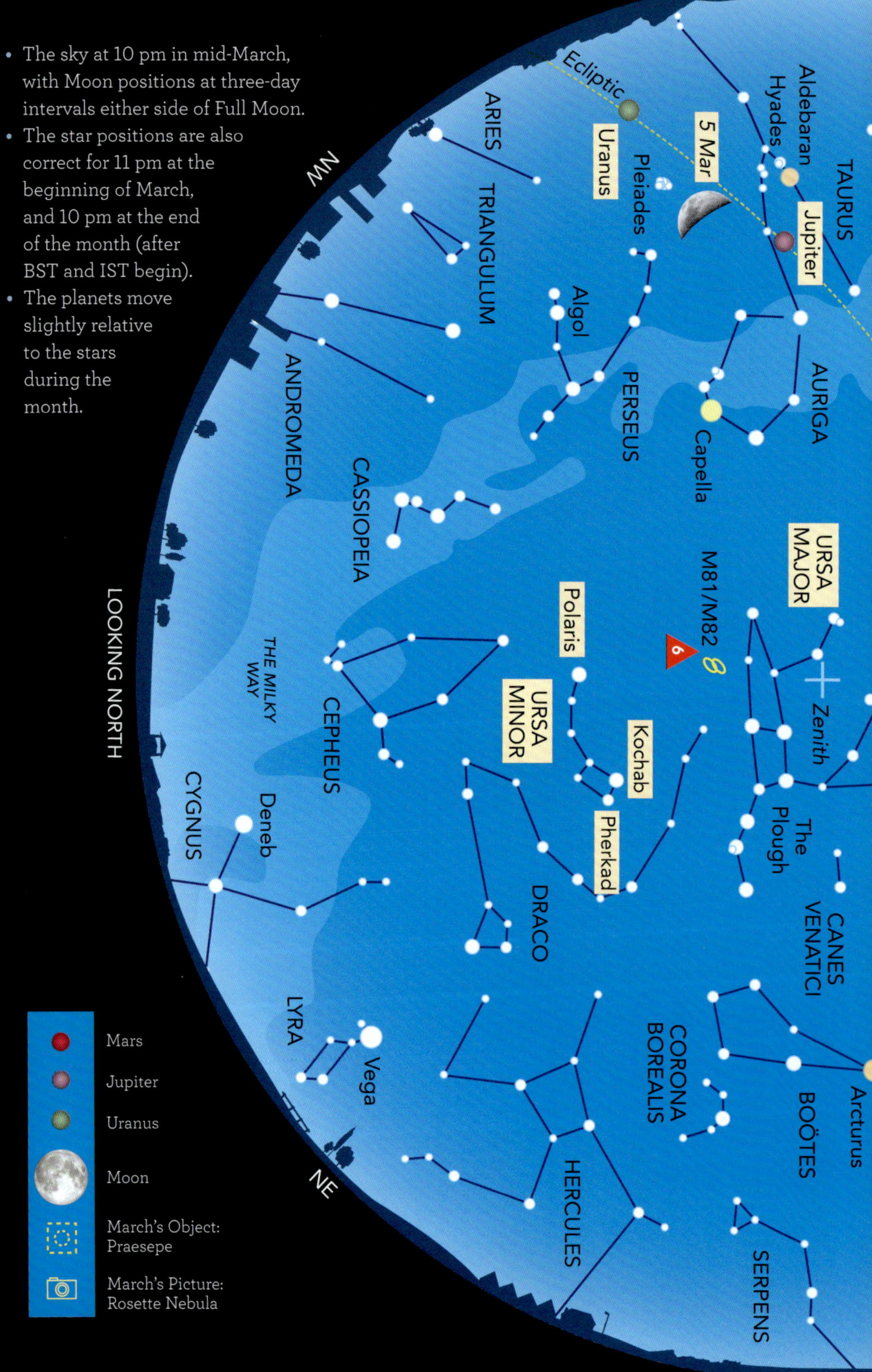

EAST
WEST
SERPENS
BOÖTES
Arcturus
CANES VENATICI
Zenith
URSA MAJOR
AURIGA
Jupiter
Aldebaran
TAURUS
Castor
8 Mar
Pollux
GEMINI
Mars
ERIDANUS
The Sickle
Praesepe
LEO
5
CANCER
Betelgeuse
VIRGO
Regulus
11 Mar
ORION
Rigel
Procyon
Rosette Nebula
CANIS MINOR
Ecliptic
Spica
14 Mar
LEPUS
THE MILKY WAY
Sirius
CORVUS
CANIS MAJOR
HYDRA
PUPPIS
SE
SW
LOOKING SOUTH
TOP 20 SKY SIGHTS
(see pp. 83–85)
5 Praesepe
6 M81 and M82

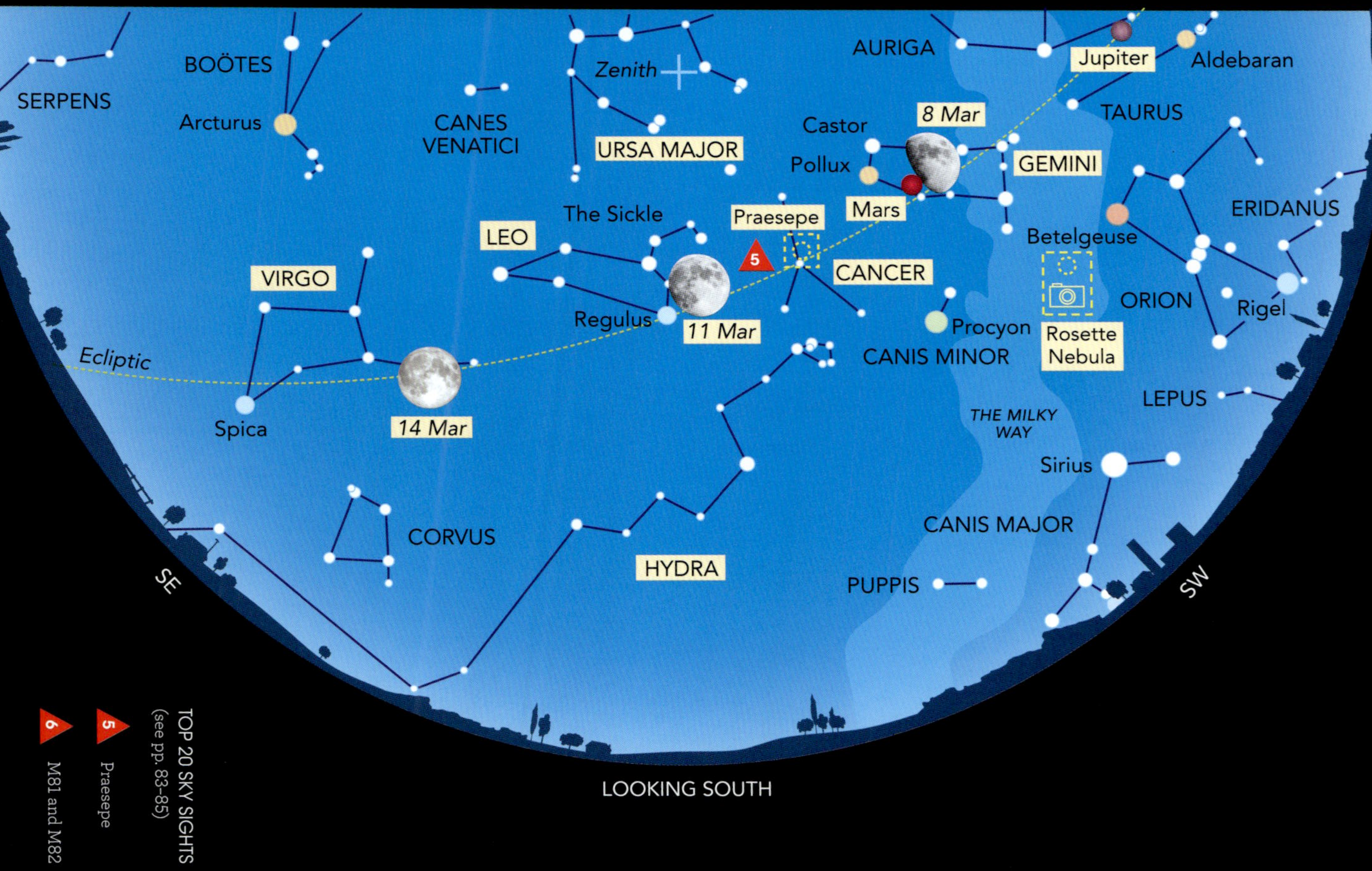

MARCH

It's eclipse month! We're treated to eclipses of both the Moon (14 March) and the Sun (29 March). After months as our Evening Star, Venus bows out this month, but we still have **Jupiter** and **Mars** to enliven the skies, along with the large spring constellations – **Leo, Virgo** and **Hydra** – now well on view in the southern sky.

MARCH'S CONSTELLATION

Ursa Minor (the Little Bear) is a miniature version of the Great Bear, **Ursa Major**, and it contains **Polaris**, the Pole Star or North Star. This star (magnitude +2.0, but slightly variable) happens to lie in line with the Earth's axis, so it appears fixed in the sky – due north – as our planet rotates. The other two bright stars in Ursa Minor are called the 'guardians of the pole': **Kochab** is an orange giant, while **Pherkad** is a blue-white star.

According to ancient Greek legend, Ursa Major was originally a beautiful nymph. When she had a son by Zeus, the great god's jealous wife turned the nymph into a bear. Years later, Zeus saw his son about to shoot this apparently wild creature. He quickly gave the young man an ursine shape, grabbed both bears by their tails and flung them into the sky.

MARCH'S OBJECT

'A nebulous mass in the breast of the Crab' is how the ancient astronomer Ptolemy described this month's object. With the newly invented telescope, Galileo scrutinised it in 1609 and found it's a cluster of more than 40 faint stars. In fact, **Praesepe** ('the manger') is a swarm of around 1000 stars, prompting the more poetic nickname the Beehive Cluster.

With the naked eye, you can spot Praesepe as a faint misty patch in **Cancer**, between **Gemini** and **Leo**. It is a magnificent sight through binoculars or a small telescope, its stars scattered like jewels on black velvet over an area three times wider than the Moon.

Lying some 600 light years away, Praesepe was born from a dense gas cloud about 600 million years ago. The cluster is 70 light years in diameter. Recent measurements from the Gaia satellite – which measures star distances and motions – reveal that members of Praesepe are leaking away into space, to form two 'tails' of stars each 540 light years long.

OBSERVING TIP

Whether you're observing the partial solar eclipse this month, or just checking out our local star, NEVER look at the Sun directly with your unprotected eyes or – especially – with a telescope or binoculars: it could blind you permanently. For naked-eye observing, use special 'eclipse glasses' with dark filters (meeting the ISO 12312-2 safety standard). Or you can attach a sheet of a specialised metallised film (such as Baader AstroSolar® film) across the front of your instrument as a safe filter.

MARCH'S TOPIC: SATURN'S RINGS

A unique and beautiful sight, whether through your telescope eyepiece or imaged in close-up by a space mission, Saturn's rings stack up the superlatives.

They are composed of over a million trillion chunks of ice orbiting the planet as 'mini moons', with sizes ranging from smaller than a pea to the dimension of a house. The shiny freshness of the rings suggests that they formed less than 100 million years ago when a small icy moon was shattered, possibly when it was hit by a comet.

Though the rings are almost wide enough to stretch from the Earth to the Moon, all the particles put together contain less ice than there is in the Antarctic ice sheet. Most astonishing of all, the rings are incredibly thin – no more than 30 metres thick. If you made a scale model out of paper, it would have to be 2 kilometres wide!

Because Saturn is tipped up, our perspective on the rings is constantly changing, and twice in every Saturn-year (30 Earth-years) we get to see the rings edge-on. They are so thin that at these 'ring-plane crossings' the rings disappear from sight. Before the era of visiting spacecraft, these were ideal times for astronomers to hunt for new moons around the sixth planet.

This month, on 23 March, there's a ring-plane crossing – but, sadly, Saturn lies so close to the Sun that we won't be able to view it. To witness the rare sight of 'Saturn without its rings' you'll have to wait until 1 April 2039.

MARCH'S PICTURE

The glorious **Rosette Nebula** is a true celestial gem, imaged here by Sara Wager under the dark skies of eastern Spain. Some 5300 light years away, the petals of the rose spread out over 130 light years of space. Violent young stars – some as brilliant as 400,000 Suns – have punched a hole in the nebula's heart. Dense clouds of gas and dust around the Rosette contain enough matter to spawn another 100,000 stars.

Sara Wager captured the Rosette Nebula with two Takahashi FSQ-85 85-mm refractors strapped together, one with a QSI 683 camera and the other with a Moravian G2-8300 camera. Using Astrodon 3-nm filters, she took multiple exposures at three wavelengths – 30 × 1800 seconds in hydrogen alpha (H-alpha), 25 × 1800 seconds in oxygen light ([OIII]) and 25 × 1800 seconds in sulphur emission ([SII]) – for a total exposure of 40 hours.

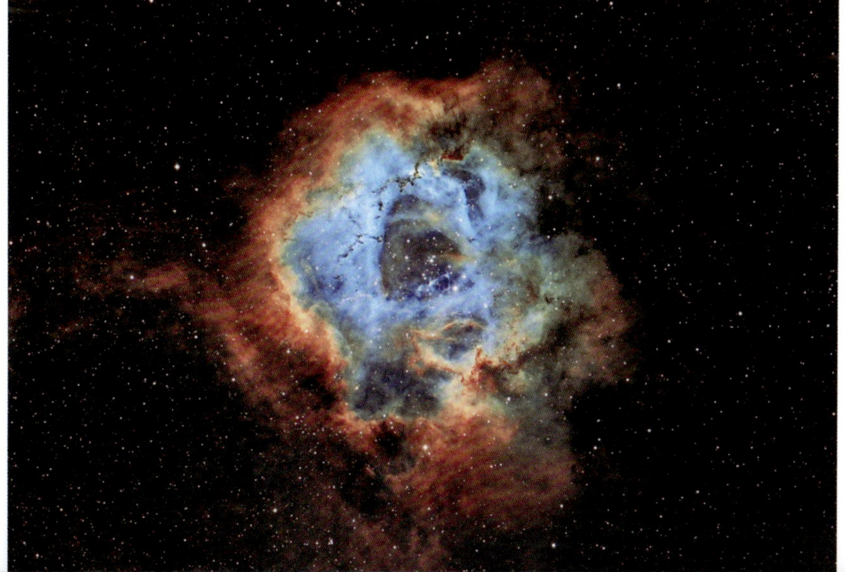

MARCH'S CALENDAR

SUNDAY	MONDAY	TUESDAY	WEDNESDAY	THURSDAY	FRIDAY	SATURDAY
30 BST and IST begin (am)	31					1 Moon near Venus and Mercury
2 Moon near Venus	3	4	5 Moon near Jupiter	6 4.31 pm First Quarter Moon near Jupiter	7	8 Mercury E elongation; Moon near Mars
9 Moon very near Mars (am)	10	11	12 Mercury near Venus	13	14 6.55 am Full Moon; total lunar eclipse	15
16	17	18	19	20 Spring Equinox	21	22 11.29 am Last Quarter Moon
23	24	25	26	27	28	29 10.58 am New Moon; partial solar eclipse

SPECIAL EVENTS

- **1 March:** the narrowest crescent Moon lies to the lower left of Venus, with Mercury below (Chart 3a).
- **2 March:** Venus forms a stunning sight with the crescent Moon (Chart 3a).
- **8 March:** Mercury reaches its greatest separation from the Sun: it's to the lower left of Venus (see Planet Watch).
- **Night of 8/9 March:** the Moon passes only 40 arcminutes away from Mars.
- **14 March:** a total eclipse of the Moon is visible from the Americas and western regions of Africa and Europe. People in Ireland and the UK will witness the partial phase beginning at 5.10 am, with totality starting at 6.26 am – just as the Moon is about to set.
- **20 March, 9.01 am:** the Spring Equinox.
- **23 March:** the Earth crosses the ring-plane of Saturn, but the ringworld is too close to the Sun for us to observe it.
- **29 March:** a partial eclipse of the Sun can be seen from parts of Europe, Africa and Canada, where 93% of the Sun is covered. For northern Scotland and Ireland, 45% of the Sun is obscured (Chart 3b), decreasing to 30% for south-east England. The partial phase starts around 10 am and ends about 12 noon with maximum eclipse between 10.55 am and 11.05 pm (exact times depend on location).
- **30 March, 1.00 am:** British Summer Time and Irish Standard Time start – put your clocks forward tonight.

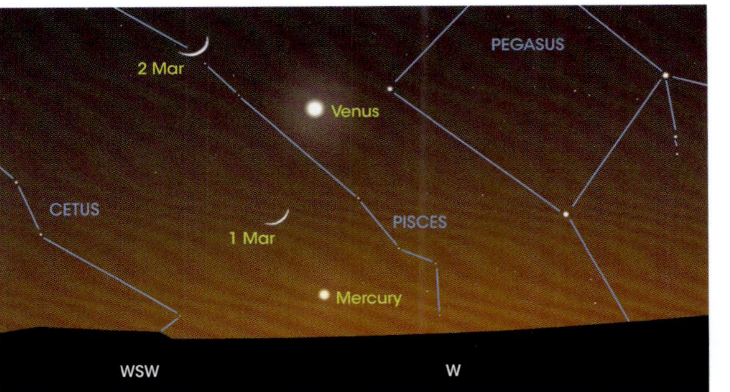

3a 1–2 March, 6.30 pm. The Moon passes Mercury and Venus.

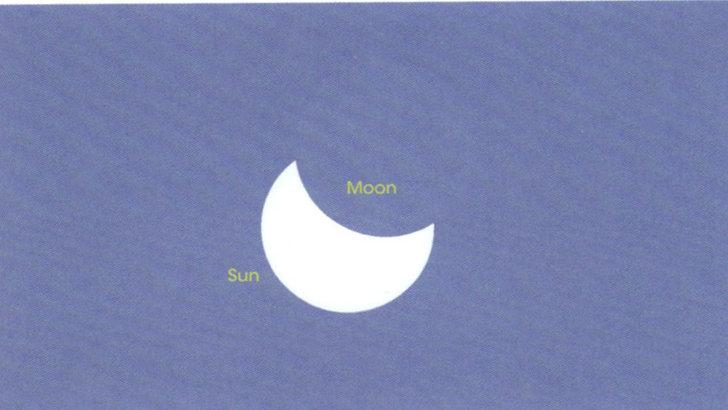

***3b** 29 March, 11.07 am (from Edinburgh). Maximum partial eclipse of the Sun.*

- At the start of March, **Venus** is shining at a brilliant magnitude –4.8 and sets just before 9 pm. On 1 and 2 March, the crescent Moon forms a pretty tableau as it passes to the left of Venus; **Mercury** (magnitude –0.9) hugs the horizon below (Chart 3a).
- As the Evening Star sinks into the dusk twilight, Mercury is moving upwards and fading. The innermost planet reaches its greatest separation from the Sun on 8 March, when it sets at 7.40 pm. By the time it passes Venus, on 12 March, Mercury has faded to magnitude +0.5.
- Although Venus has been a gleaming jewel in the evening sky since December, this month it swings between the Earth and the Sun and disappears from the evening sky. On the last few days of March, you may catch sight of Venus low in the east before dawn as the Morning Star.
- **Jupiter** lies in Taurus at magnitude –2.2, and sets around 1.30 am. The Moon is nearby on 5 and 6 March.
- Lying in Gemini, **Mars** shines at magnitude +0.1 and sets about 4.30 am. The Moon passes close by on the night of 8/9 March (see Special Events).
- **Uranus** (magnitude +5.8) moves from Aries to Taurus this month, and is setting around 11.30 pm.
- **Saturn** and **Neptune** are too close to the Sun to be seen in March.

- The sky at 11 pm in mid-April, with Moon positions at three-day intervals either side of Full Moon.
- The star positions are also correct for midnight at the beginning of April, and 10 pm at the end of the month.
- The planets move slightly relative to the stars during the month.

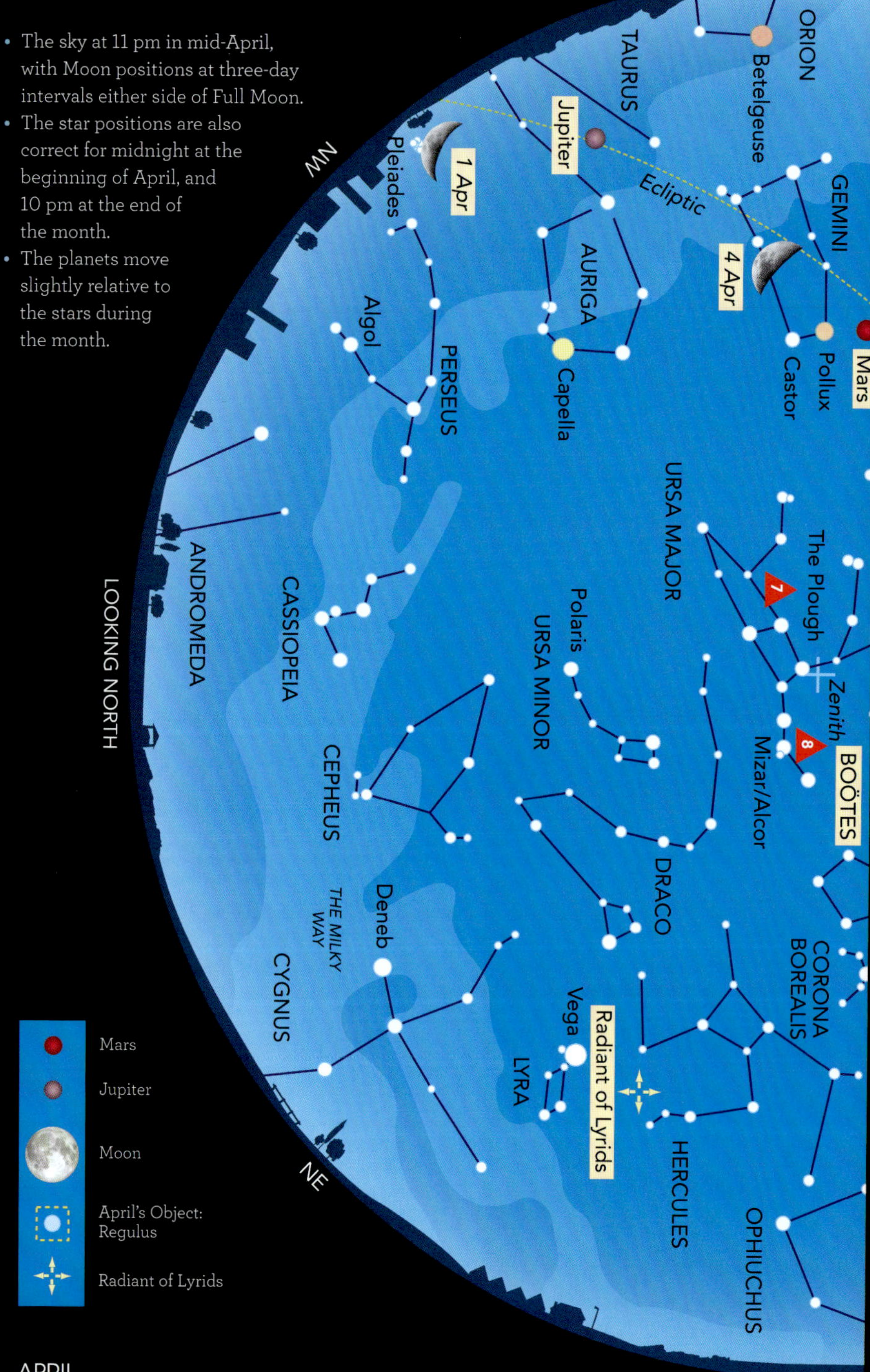

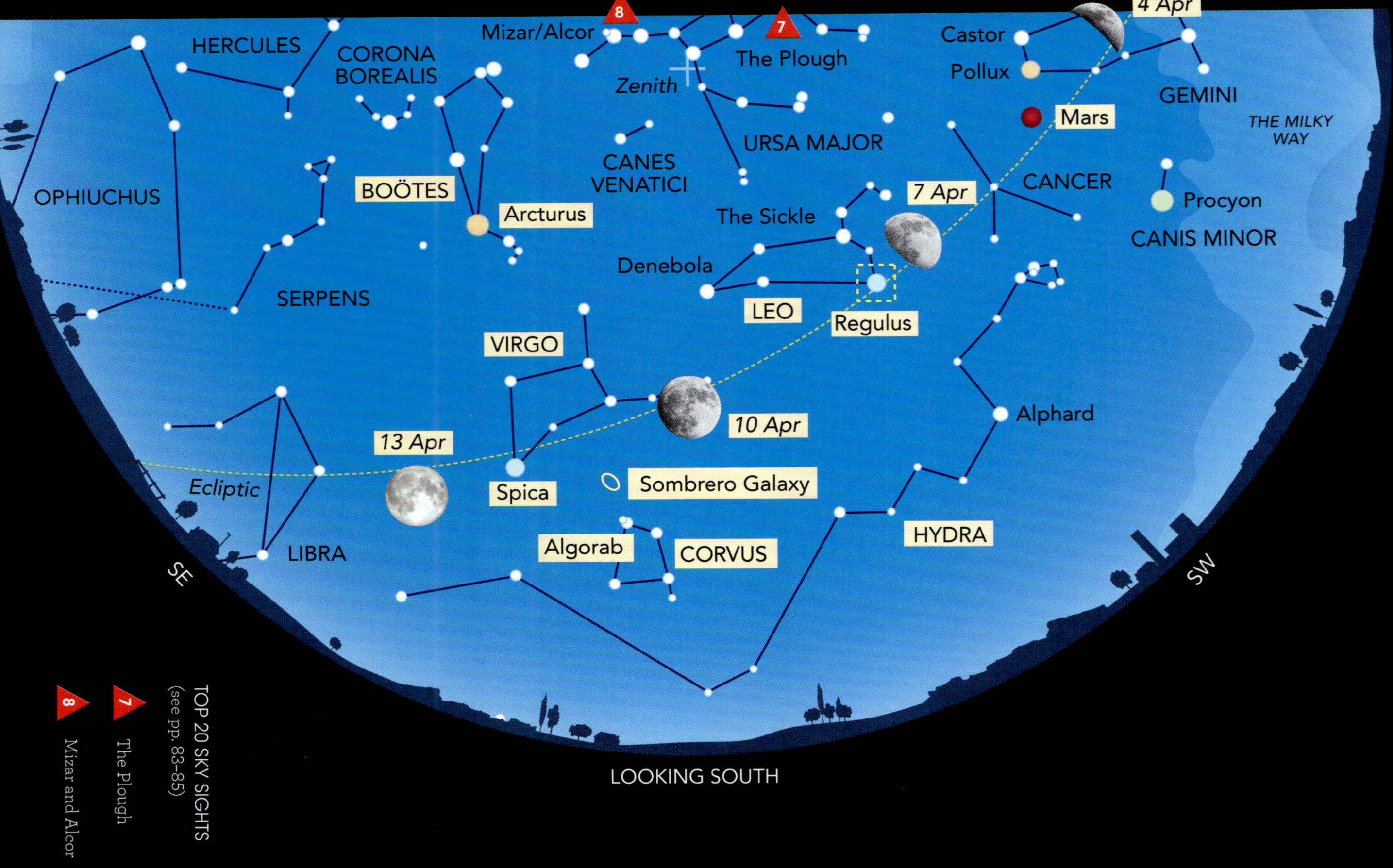

APRIL

TOP 20 SKY SIGHTS
(see pp. 83–85)

- 7 The Plough
- 8 Mizar and Alcor

Watch out for the Moon hiding the Seven Sisters and a shower of shooting stars this month. The brilliant planets **Jupiter** and **Mars** in the west are counterbalanced by three bright stars to the south-east: **Regulus** in **Leo**, **Virgo**'s leading star **Spica** to its lower left, and orange **Arcturus** in **Boötes** lying above.

APRIL'S CONSTELLATION

Corvus (the Crow) is a distinctive quadrilateral of stars that stands out in a rather barren region of sky. The early Babylonians saw it as a raven, while to the Greeks this star pattern was a crow, the bird sacred to the god Apollo.

In Greek legend, Corvus was originally white, but when it informed Apollo that his lover had been unfaithful, the god – in a furious rage – turned the bird black.

Later, Apollo commanded the crow to fetch water; but Corvus tarried to feast on the ripe fruit of a fig tree by the spring. The bird picked up a water snake which – it said – had caused a delay by obstructing the spring. Seeing through the excuse, Apollo flung the bird and snake skywards, as Corvus and neighbouring **Hydra**.

Algorab is a double star – white and orange – that you can split in a small telescope. You'll find the **Sombrero Galaxy** on the border of Corvus and **Virgo**. Just visible in binoculars, you'll need a moderate telescope to discern the dark dust streaks making up the brim of this celestial Mexican hat.

APRIL'S OBJECT

Regulus – the 'heart' of **Leo** the Lion – appears to be a bright but run-of-the-mill star. Some 79 light years away, it's young (about 250 million years old), 3.4 times heavier than the Sun, and 360 times more luminous than our local star.

From the coast of the Beara Peninsula, County Cork, Ireland, Denis Walsh was watching the Sun setting near the Skellig Rocks out in the Atlantic when he witnessed this glorious sight. He captured it with a Nikon D40X camera attached to a Celestron 102-mm f/5 refractor, with an exposure of 1/2500 second at ISO 100.

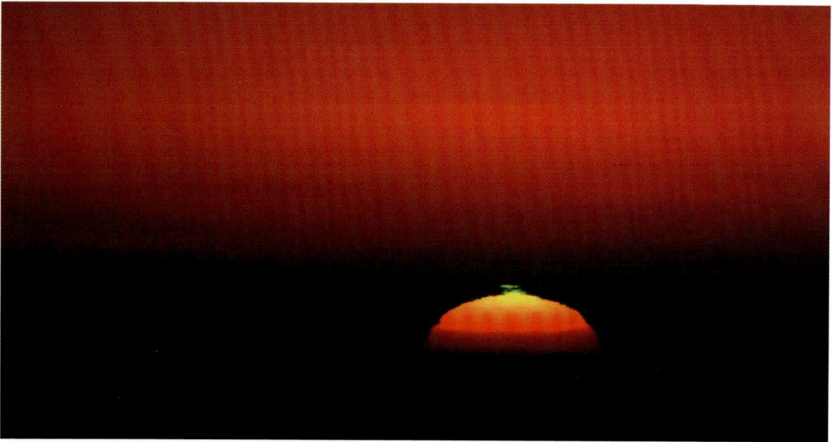

But detailed observations reveal Regulus to be a whirling dervish, spinning round in just 16 hours (as compared to roughly a month for our Sun). As a result, its equator is rotating at a million kilometres per hour, and bulges outwards to give the star a tangerine shape. If Regulus span round only 5 per cent faster, the star would tear itself apart!

Regulus has been spun-up by grabbing gas from a closely orbiting star, which has now been reduced to a shrunken white dwarf. There's a more distant companion that you can spot with even a small telescope: a 500-mm instrument shows you that this star is also double, making Regulus a four-star system.

APRIL'S TOPIC: DATE OF EASTER

The moveable feast of eggs and chocolate falls on 20 April this year, a long time after most other spring celebrations, and near the end of the possible range of dates that spans 22 March–25 April. If you've ever wondered why the date of Easter changes so much, it's all to do with astronomy . . .

According to the Bible, Jesus was crucified at the Passover, whose date was fixed by a calendar based on the Spring Equinox and the Moon's phases. So you can tell when Easter is due just by looking at the sky. First, wait until you see the Sun rising due east and setting due west: that's the Spring Equinox. Now follow the Moon until it's Full – and Easter will be the next Sunday. This year, the Full Moon after the equinox falls on 13 April, and the following Sunday is 20 April.

There's an even later Easter on 21 April 2030; while in 2038 your choccy eggs will have to wait until 25 April – a date that occurs only once per century.

OBSERVING TIP

It's best to view your favourite objects when they're well clear of the horizon. If you observe them low down, you're looking through a large thickness of the atmosphere – which is always shifting and turbulent. It's like trying to observe the outside world from the bottom of a swimming pool! This turbulence makes the stars appear to twinkle. Low-down planets also twinkle – although to a lesser extent, because they subtend tiny discs, and aren't so affected.

APRIL'S PICTURE

If you're surprised to see the top of the setting Sun suddenly gleam like a brilliant emerald, you're not hallucinating! Instead you're witnessing a rare 'green flash'.

When the Sun begins to set, the Earth's atmosphere bends its rays, diverting each wavelength (colour) to a slightly different extent. As a result, the top of the Sun's disc becomes a spectrum of colours, fringed with green at the top (there's a blue layer above the green, but the blue light is usually scattered away by the atmosphere).

If the air lies in stable layers, the green fringe can separate from the solar disc, as Denis Walsh witnessed from the Irish coast. A green flash occurs when the main body of the Sun has set, and we see for a moment just this intense viridescent glint.

Once, when visiting the high-altitude observatory on Mount Teide in the Canary Islands, I was lucky enough to witness an even more exceptional sight. The air was too thin to scatter away the blue component of sunlight. As the Sun set behind the mountain's volcanic flank, its last speck gleamed out as the intense sapphire of a 'blue flash'.

APRIL'S CALENDAR

SUNDAY	MONDAY	TUESDAY	WEDNESDAY	THURSDAY	FRIDAY	SATURDAY
		1 Moon occults Pleiades	2 Moon occults Pleiades (am); Moon near Jupiter	3	4	5 3.15 am First Quarter Moon near Mars
6	7	8	9	10	11	12 Moon near Spica
13 1.22 am Full Moon near Spica (am)	14	15	16	17 Moon near Antares (am)	18	19
20	21 2.35 am Last Quarter Moon; Mercury W elongation	22 Venus at greatest brightness (am); Lyrids	23 Lyrids (am)	24 Moon near Venus (am)	25 Moon near Venus (am)	26
27 8.31 pm New Moon	28	29	30 Moon near Jupiter			

SPECIAL EVENTS

- **Night of 1/2 April, 9.30 pm–0.30 am:** the Moon passes right in front of the Pleiades, hiding in turn six of the Seven Sisters' brightest stars (Chart 4a). Find an observing site with a clear north-western horizon, because the Moon is low in the sky and sets as the occultation is ending.
- **2 April:** the crescent Moon passes Jupiter.
- **5 April:** the reddish 'star' near the First Quarter Moon is Mars (Chart 4b).
- **20 April:** NASA's Lucy mission, en route to the Trojan asteroids that follow Jupiter's orbit, flies by the carbon-rich asteroid Donaldjohanson.
- **21 April:** Mercury is at its greatest separation from the Sun, but lost in the twilight glow.
- **22 April, before dawn:** Venus reaches its maximum brightness (see Planet Watch).
- **Night of 22/23 April:** it's a great year for observing the maximum of the **Lyrid meteor shower** against a dark sky as the Moon doesn't rise until just before dawn. These shooting stars – debris from Comet Thatcher – often leave dusty trails as they burn up in the Earth's atmosphere.
- **24 April, before dawn:** the crescent Moon lies to the right of Venus.
- **25 April, before dawn:** Venus is just above the crescent Moon.
- **30 April:** Jupiter lies below the crescent Moon.

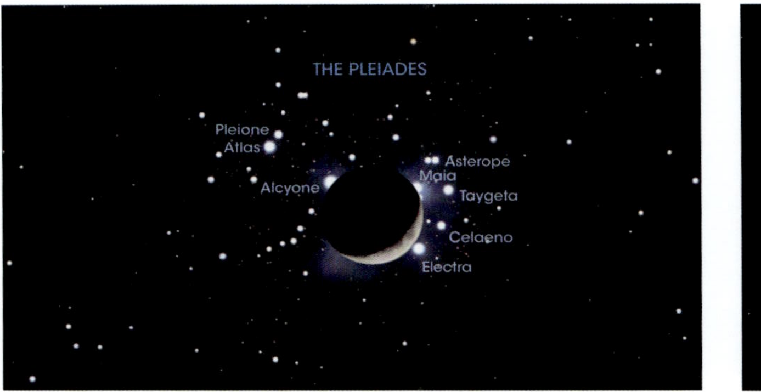

4a *1 April, 10.45 pm. The Moon occults the Pleiades.*

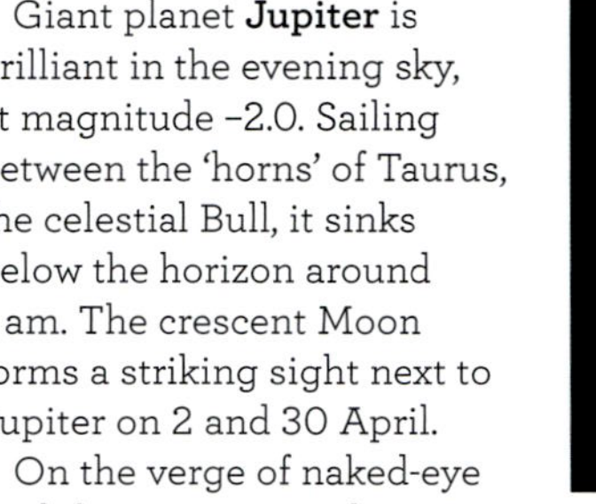

4b *5 April, 10 pm. The Moon joins Mars, Pollux and Castor.*

• Giant planet **Jupiter** is brilliant in the evening sky, at magnitude –2.0. Sailing between the 'horns' of Taurus, the celestial Bull, it sinks below the horizon around 1 am. The crescent Moon forms a striking sight next to Jupiter on 2 and 30 April.

• On the verge of naked-eye visibility at magnitude +5.8, **Uranus** lies in Taurus, below the Pleiades. Setting about 10.30 pm, the seventh planet is lost in the twilight glow by the end of the month.

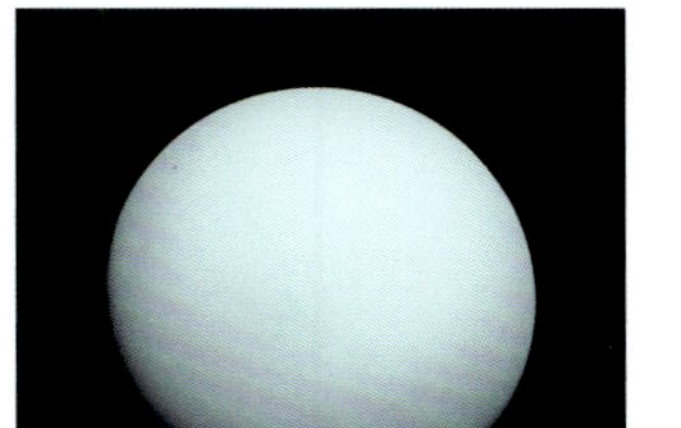
Uranus

• **Mars** starts this month in Gemini, forming a triangle with Castor and Pollux, and – at magnitude +0.6 – slightly outshining the twin stars. The Red Planet sets around 3.30 am. The Moon passes near Mars on 5 April (Chart 4b). On 10 April, Mars, Pollux and Castor are in line. By the end of April, Mars has moved into Cancer.

• Low in the morning twilight, **Venus** rises about 5 am. It reaches its maximum brilliance as a Morning Star (magnitude –4.8) on 22 April. You'll find the crescent Moon near Venus on the mornings of 24 and 25 April.

• **Saturn**, **Neptune** and **Mercury** are lost in the Sun's glare this month, even though Mercury reaches its greatest separation from the Sun on 21 April.

APRIL'S PLANET WATCH

- The sky at 11 pm in mid-May, with Moon positions at three-day intervals either side of Full Moon.
- The star positions are also correct for midnight at the beginning of May, and 10 pm at the end of the month.
- The planets move slightly relative to the stars during the month.

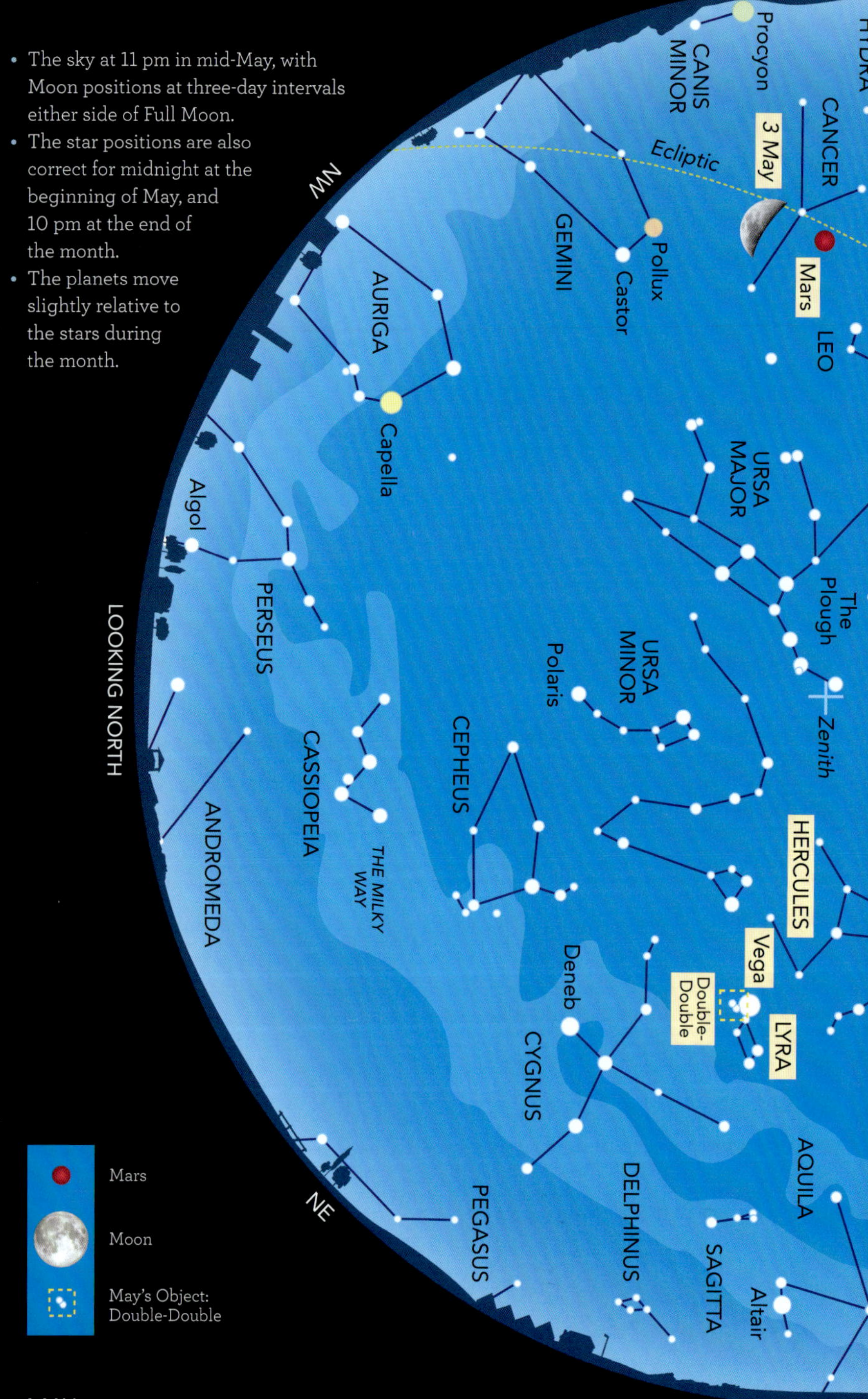

MAY

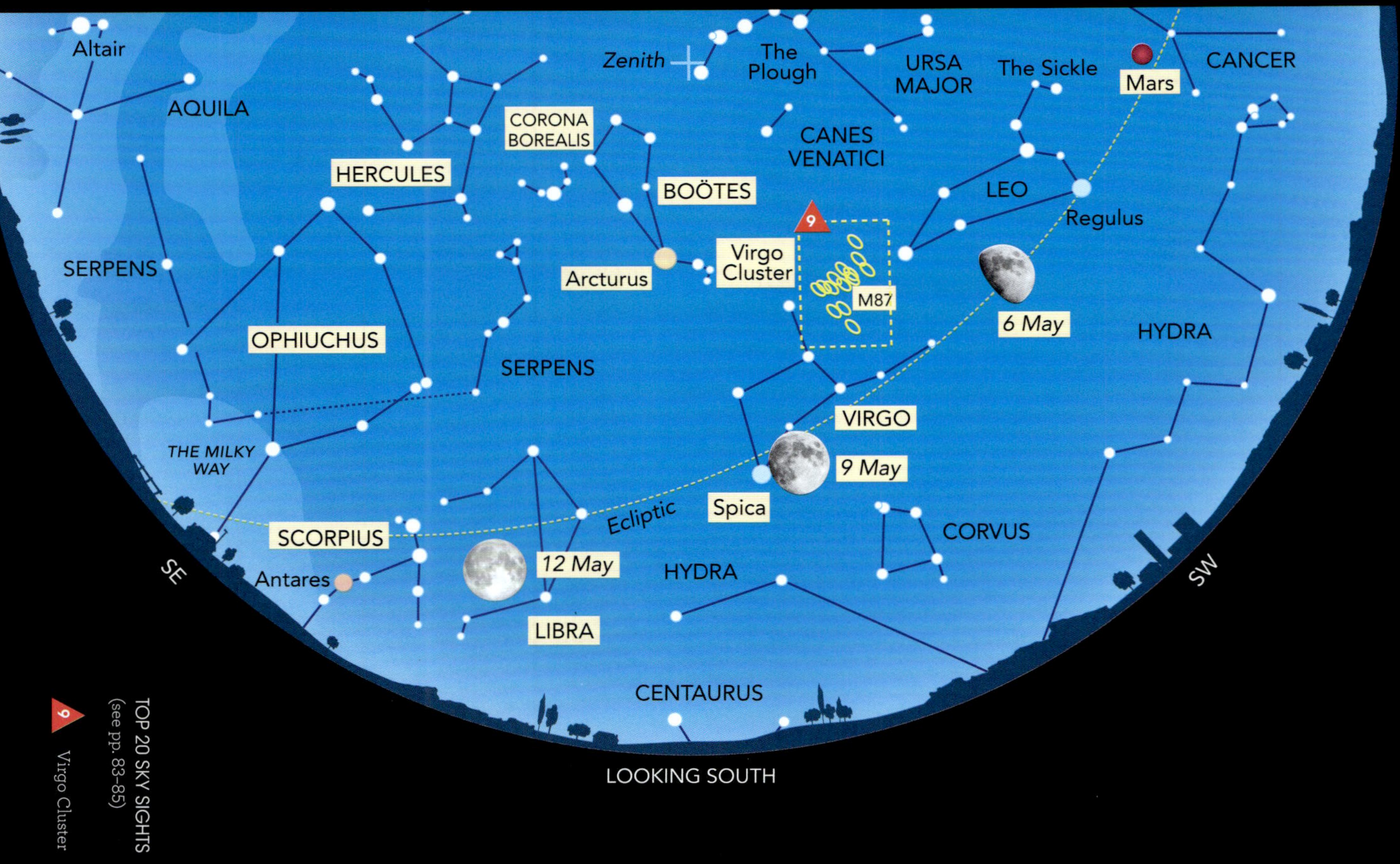

TOP 20 SKY SIGHTS
(see pp. 83–85)

9 Virgo Cluster

The night is bookended by Jupiter and **Mars** visible after sunset, and Venus and Saturn presaging the dawn. And enjoy making the acquaintance of the early summer constellations: **Boötes**, with the little circlet of **Corona Borealis** to its left; **Scorpius**, **Libra** and **Virgo** to the south; and two faint sprawling giants – **Hercules** and **Ophiuchus** – in the south-eastern sky.

MAY'S CONSTELLATION

The Y-shaped constellation of **Virgo** is the second largest in the sky. To the Greeks, she was the chaste goddess of justice, with her scales lying next to her as the neighbouring constellation **Libra**. But previously these stars were seen as a fertile furrow in a farmer's field, and the name **Spica** ('ear of corn') for the brightest star reflects this ancient Babylonian tradition.

Spica is a hot, blue-white star over 20,000 times brighter than the Sun, boasting a temperature of 25,000°C. It has a stellar companion, which lies closer than Mercury orbits the Sun. The stars inflict a mighty gravitational toll on each other, stretching each star into an egg shape.

The glory of Virgo lies in the 'bowl' of the Y. Scan the upper region with a small telescope – at a low magnification – and you'll find it packed with faint, fuzzy blobs. These are just a few of the 2000 galaxies that make up the gigantic **Virgo Cluster**. It's centred on the mammoth galaxy **M87**, which boasts a central black hole over 6 billion times heavier than the Sun: it was the first black hole to be imaged, in 2019.

MAY'S OBJECT

Epsilon Lyrae, right next to brilliant **Vega**, is a celestial eyesight test chart. If you are young, with 20/20 vision, you'll easily discern this star to be double, with components almost equal in brightness. But anyone over 60 will struggle to see epsilon Lyrae as more than just a single star.

Grab a pair of binoculars, and you'll easily separate the stars. And a good telescope reveals that each member is itself a closely matched pair of stars – hence epsilon Lyrae's nickname, the **Double-Double**. These close pairs both have separations of about 2.5 arcseconds, around one-hundredth of the distance between the two pairs.

The four stars are all pretty similar to Sirius, about twice as massive as the Sun. This lovely set of cosmic quadruplets is a billion years old, and lies 160 light years away.

MAY'S TOPIC: GALAXY CLUSTERS

Galaxies are gregarious. Thanks to gravity, they like living in groups. The Milky Way and our neighbouring giant spiral the Andromeda Galaxy are members of a small cluster of about 30 smallish galaxies called the Local Group.

But the Virgo Cluster (see Constellation) is in a different league. It's our nearest giant cluster of galaxies, lying at a distance of 54 million light years. As well as keeping its own myriad galaxies in check, the powerful gravity of the Virgo Cluster holds sway over the smaller groups around – including our Local Group – making up a cluster of clusters of galaxies, the Virgo Supercluster.

The Universe teems with a billion galaxy clusters, some of them ten times heavier than the Virgo Cluster. In their crowded centres, collisions are frequent, converting stately spiral galaxies into bland ellipticals (see January's Topic). And the clusters generally bond with their neighbours in gargantuan superclusters.

Galaxy clusters and superclusters are the building blocks of the Universe. They make up the structure of the Cosmos, with huge filaments of superclusters enclosing enormous empty voids, like a gigantic Swiss cheese.

MAY'S PICTURE

'Green comet greets the Red Planet' would be a fitting title for this memorable image taken by Martin Ratcliffe from the mountains of Utah, USA. The planet was Mars, some two months after passing closest to Earth in December 2022. Its companion was Comet ZTF, named after the telescope used to find it, the Zwicky Transient Facility (see February's Topic for the dedicatee).

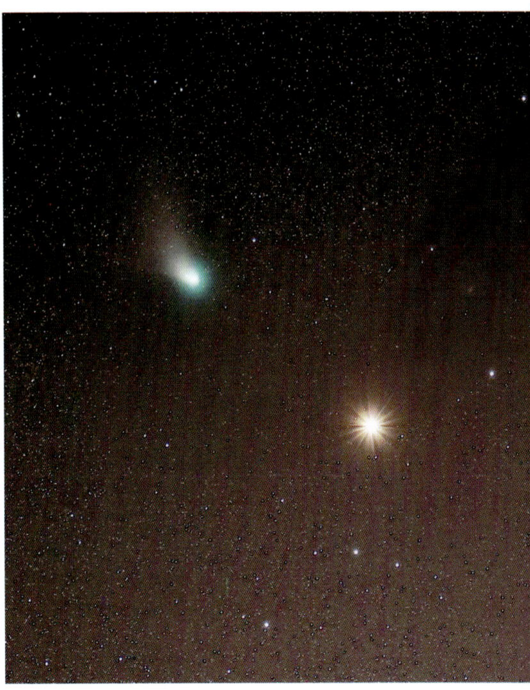

'Finally, a clear sky and no moon,' enthused Martin Ratcliffe as he reached a 2000-m high mountain pass west of Rush Valley, Utah. Setting up his Canon 6D Mark II camera with a 135-mm f/2 Rokinon lens, he bagged this shot on 11 February 2023 at 3.09 am GMT. It was a 2-minute exposure at ISO 1000.

Comet ZTF had passed the Sun a month earlier and was now heading outwards to the farthest reaches of the Solar System. Although the comet looks as though it's narrowly missing Mars, it's just a line-of-sight effect, as the Red Planet was then over twice as far away as Comet ZTF.

And the colour contrast? Mars is red because its desert sands have literally rusted. The comet's green light is emitted by carbon molecules excited by the Sun's ultraviolet radiation.

OBSERVING TIP

When you first go out to observe, you may be disappointed at how few stars you can see in the sky. But wait for 20 minutes, and you'll be amazed at how your night vision improves. One reason for this 'dark adaptation' is that the pupil of your eye grows larger. More importantly, in dark conditions the retina of your eye builds up bigger reserves of rhodopsin, the chemical that responds to light.

MAY'S CALENDAR

SUNDAY	MONDAY	TUESDAY	WEDNESDAY	THURSDAY	FRIDAY	SATURDAY
				1	2 Moon near Castor and Pollux	3 Moon near Mars and Praesepe
4 2.52 pm First Quarter Moon; Mars very near Praesepe	5 Moon near Regulus; Mars near Praesepe	6 Eta Aquarids (am)	7	8	9 Moon near Spica	10
11	12 5.56 pm Full Moon	13	14 Moon near Antares (am)	15	16	17
18	19	20	21 0.59 am Last Quarter Moon	22 Moon near Saturn (am)	23 Moon between Venus and Saturn (am)	24 Moon near Venus (am)
25	26	27 4.02 am New Moon	28 Moon near Jupiter	29	30	31 Moon near Mars

SPECIAL EVENTS

- **3 May:** the Moon passes over Mars; look carefully (a pair of binoculars will help) and you'll spot the star cluster Praesepe nestled up close to Mars as well (Chart 5a).
- **3–5 May:** Mars moves through the upper fringes of Praesepe in Cancer.

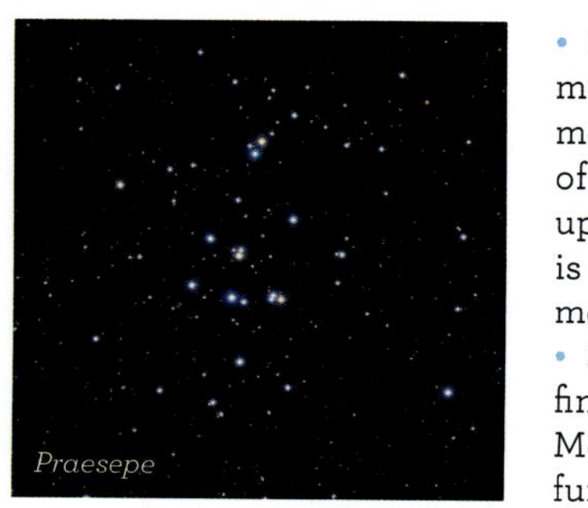
Praesepe

- **6 May, early hours:** the maximum of the Eta Aquarid meteor shower – tiny pieces of Halley's Comet burning up in Earth's atmosphere – is spoilt this year by bright moonlight.
- **22 May, before dawn:** you'll find Saturn to the left of the Moon, with brilliant Venus further left again (Chart 5b).
- **23 May, before dawn:** the Moon lies between Venus and fainter Saturn (Chart 5b).
- **24 May, before dawn:** the narrow crescent Moon forms a lovely sight with Venus in the morning twilight, with Saturn to the right (Chart 5b).
- **28 May:** Jupiter lies directly below the thin crescent Moon.

MAY'S PLANET WATCH

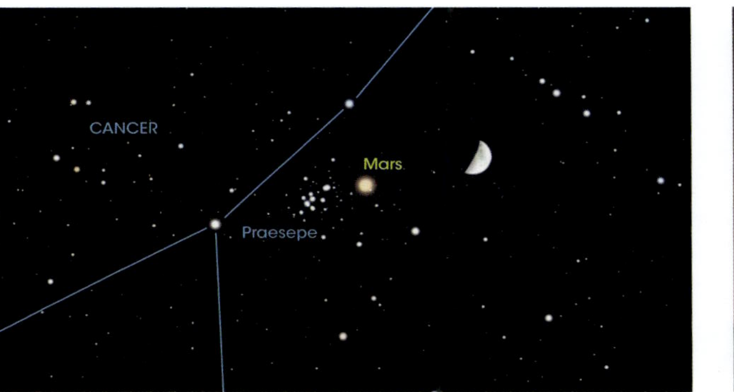

5a *3 May, 11 pm. The Moon and Mars close to Praesepe.*

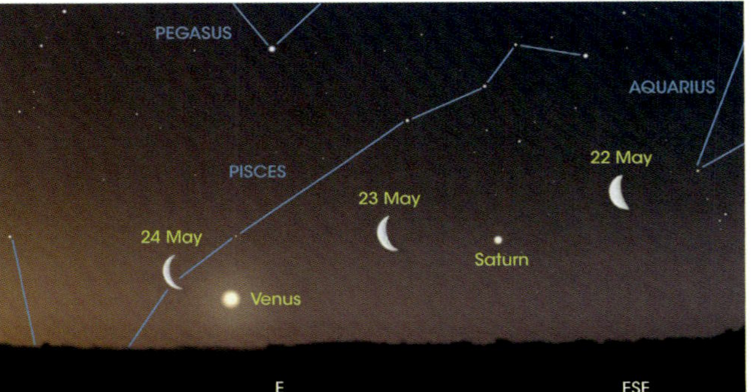

5b *22–24 May, 4 am. The Moon passes Saturn and Venus.*

- Brilliant **Jupiter** lies low in the north-west. It's fading as the Earth pulls away from the giant planet, but at magnitude –1.9 Jupiter is still more brilliant than any of the stars. The king of the planets lies in Taurus and sets about 11 pm: by the end of May, it's sinking into the evening twilight. You may just catch a thin crescent Moon above Jupiter on 28 May.
- **Mars** is a reddish 'star' in Cancer, between Regulus in Leo and the twin stars of Gemini: Castor and Pollux. Shining at magnitude +1.1, the Red Planet sets around 2 am. On the nights of 3–5 May, Mars skims the top of the star cluster Praesepe, with the Moon in attendance on 3 May (see Special Events and Chart 5a). The crescent Moon lies to the right of Mars again on 31 May.
- Rising around 4 am, **Venus** is a brilliant Morning Star (magnitude –4.6) low on the eastern horizon before dawn. The Moon is nearby on the morning of 24 May (see Special Events and Chart 5b).
- Also clearing the dawn horizon about 4 am, **Saturn** lies to the lower right of Venus at the start of May. During the month it moves to the right in the sky, rising earlier and becoming more visible against a darker sky. At magnitude +1.1, the ringworld lies in Pisces. The Moon is in the vicinity of Saturn on the mornings of 22 and 23 May (Chart 5b).
- Turn a telescope to Saturn, and you'll see it has changed its appearance since it was visible in the evening sky at the start of 2025. In March, the Earth passed through the plane of Saturn's rings, so instead of seeing these appendages from the top, we are now looking up at the rings from below.
- **Mercury**, **Uranus** and **Neptune** lie too near the Sun to be visible this month.

- The sky at 11 pm in mid-June, with Moon positions at three-day intervals either side of Full Moon.
- The star positions are also correct for midnight at the beginning of June, and 10 pm at the end of the month.
- The planets move slightly relative to the stars during the month.

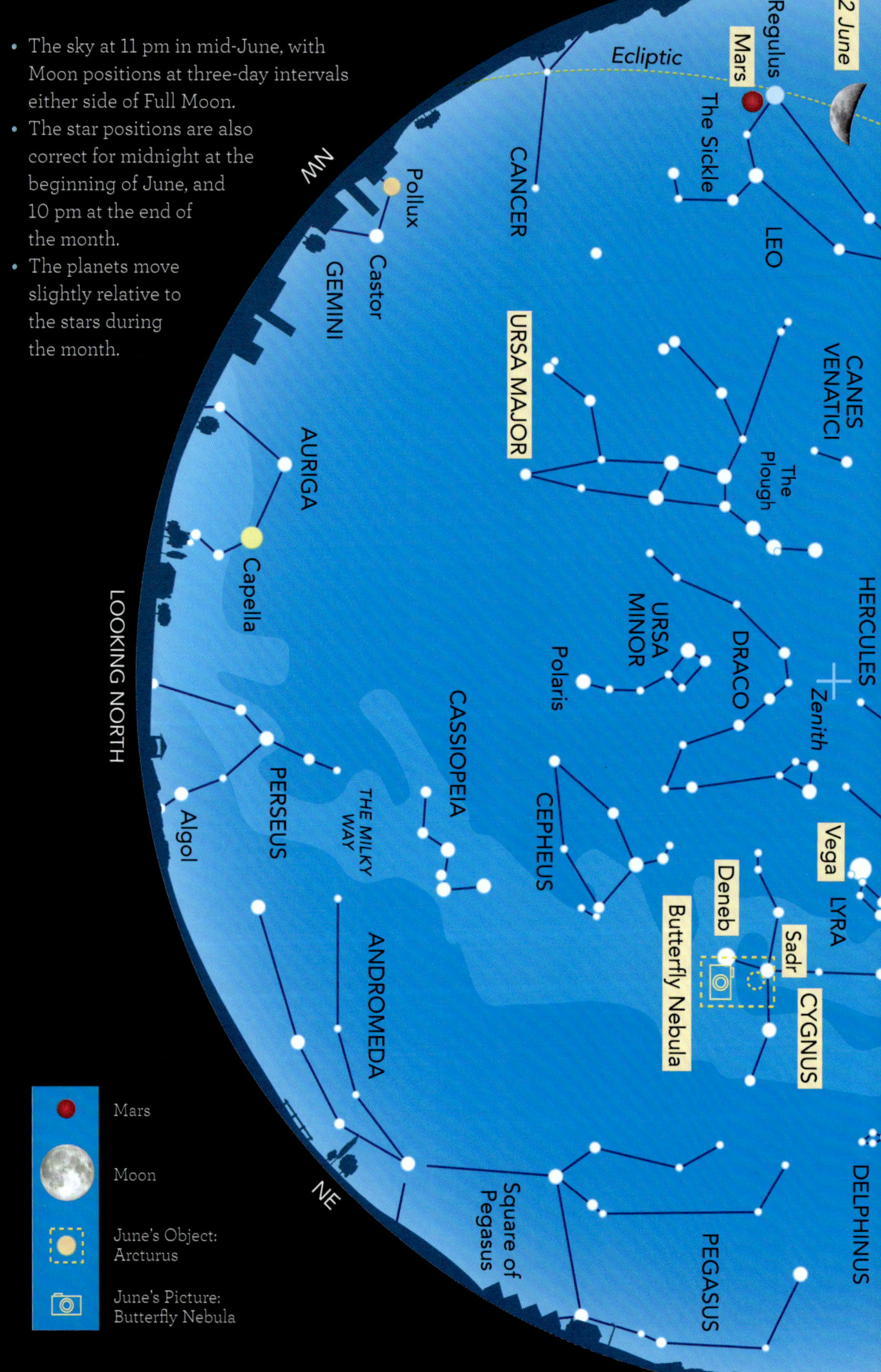

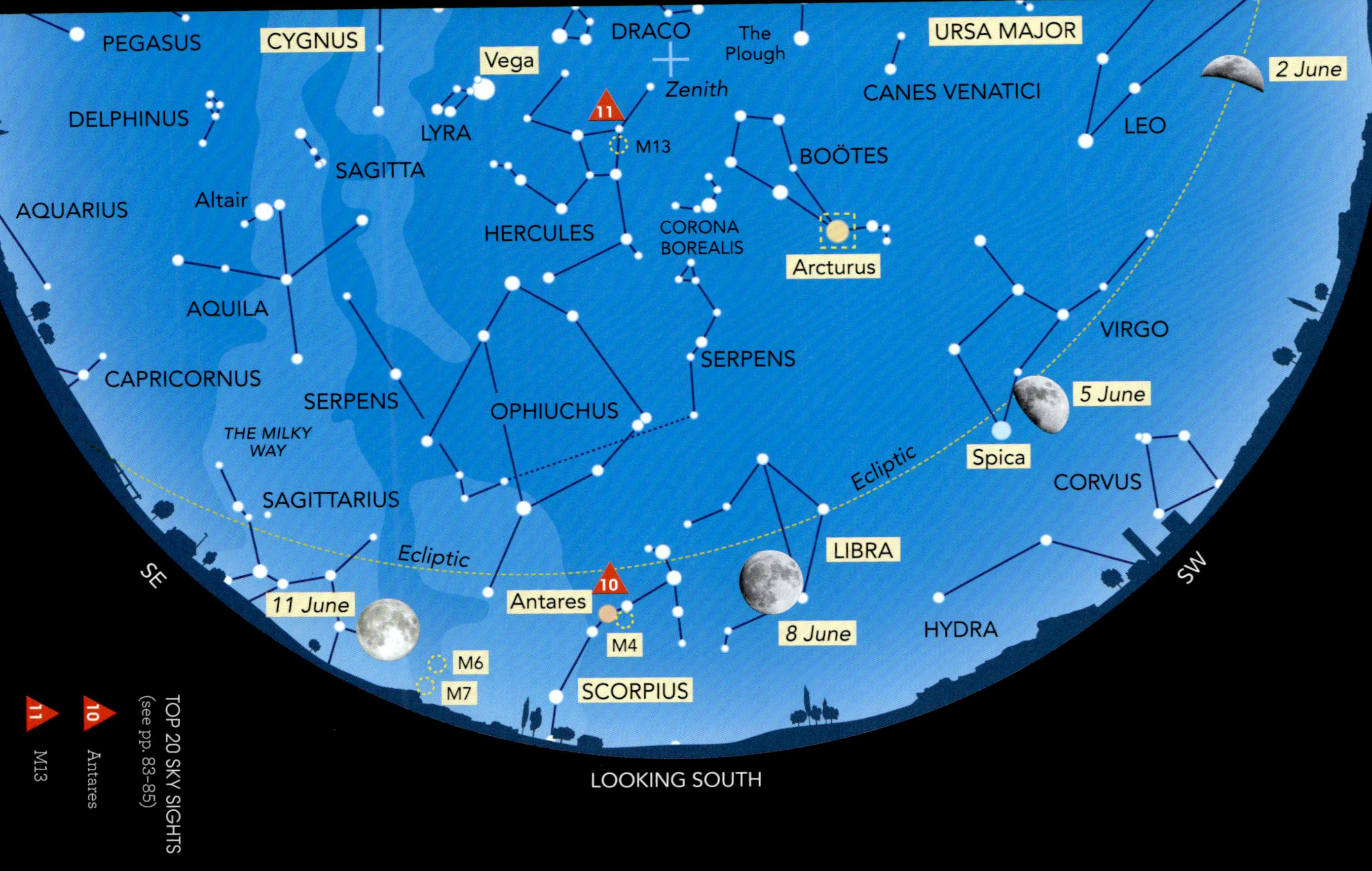

TOP 20 SKY SIGHTS
(see pp. 83–85)

10 Antares

11 M13

JUNE

This month, we're treated to the highest Sun of the year, and the lowest Full Moon for 19 years. With a paucity of bright planets in the evening sky, spend some time getting acquainted with the A-lister stars this month, including **Arcturus**, **Spica**, **Vega**, **Deneb** and **Antares**.

JUNE'S CONSTELLATION

Down in the deep south lies a baleful red star. To the ancient Greeks **Antares** – 'the rival of Mars' – marked the heart of **Scorpius**, the scorpion that killed the great hunter Orion (see January's Constellation).

Scorpius is one of the few constellations that resembles its namesake. A line of stars at the top right marks the scorpion's forelimbs. Originally, the stars we now call **Libra** (the Scales) were its claws. Below Antares, the scorpion's body stretches down into a fine curved tail (below the horizon on the Star Chart) and deadly sting.

Several lovely double stars include Antares, its faint companion appearing greenish in contrast to Antares's strong red hue. Binoculars reveal the fuzzy patch of **M4**, a globular cluster made of tens of thousands of stars; at a distance of 6000 light years, it's our nearest globular cluster. Above the 'sting' lie two fine star clusters – **M6** and **M7** – visible to the naked eye when they're well above the horizon: a telescope reveals their stars clearly.

OBSERVING TIP

Don't think that you need a telescope to bring the heavens closer. Binoculars are excellent – and you can fling them into the back of the car at the last minute. For astronomy, buy binoculars with large lenses coupled with a modest magnification. An ideal size is 7 × 50, meaning that the magnification is seven times, and that the diameter of the lenses is 50 millimetres.

JUNE'S OBJECT

The fourth-brightest star in the heavens, orange-coloured **Arcturus** was a navigational beacon for Polynesian sailors because it passes directly over Hawai'i: they called it *Hōkūle'a* ('star of joy'). The Greek name means 'the bear guardian,' as Arcturus always follows **Ursa Major** (the Great Bear) around the sky.

The brightest star in Boötes (the Herdsman), Arcturus has about the same mass as the Sun. But it's significantly older, and provides us with a preview of our local star's future. On its way to becoming a red giant, Arcturus has grown to 25 times the Sun's width and is 200 times more luminous. When the Sun reaches Arcturus's age, life on Earth will be untenable – but don't hold your breath, as that's still 2.5 billion years in the future!

JUNE'S TOPIC: LUNAR STANDSTILLS

'The Moon in June' has to be the most romantic astronomical sight, with amorous couples promenading on a warm summer's night under a Full Moon hanging low in the sky. And in 2025, the Moon seems lower and closer than ever . . .

Every year the Full Moon is at its minimum altitude in June, because it lies opposite to the Sun – which is at its highest point this month. But the Moon's orbit is tipped up, so it can travel above and below the Sun's path in the sky, the Ecliptic.

Once every 19 years, the Moon lies furthest south of the Ecliptic at the point where the Ecliptic itself is at its most southerly (in Sagittarius), and so we witness the lowest possible Full Moon. Astronomers call this point a 'major standstill', as the Moon stops moving south and then moves up in the sky again.

Not only does the Full Moon this month merely skim the horizon; the positions of moonrise and moonset are at their closest together. Neolithic people may have set up some of their standing stones in line with these important directions. Sight along the stones of Callanish in the Outer Hebrides – 'the Stonehenge of the North' – and you are looking directly at the point where the Full Moon sets at a major standstill. Unlike the solsticial sunrises and sunsets at Stonehenge, which occur every year, you'll have to wait 19 years to see this unusual and striking event again.

JUNE'S PICTURE

The ethereal **Butterfly Nebula** sits at the heart of another flying creature, **Cygnus**, the celestial Swan. To the right in Peter Jenkins's stunning image is the foreground star **Sadr**, the central star in Cygnus, whose Arabic name means the 'hen's breast' – possibly because swans are rare in the Middle Eastern deserts!

The Butterfly Nebula – formally known as IC 1318 – flutters about 5000 light years away from us. Four times the width of the Full Moon as seen in our skies, its wings are a single giant cloud of glowing gas, with delicate traceries defined by narrow bands of dark dust. Silhouetted in front, a larger cloud of dust forms the butterfly's body.

Under Nottinghamshire skies, Peter Jenkins netted the heavenly butterfly with an Officina Stellare 115-mm refractor and an Atik Horizon Mono CMOS camera. For this two-panel mosaic, Peter used narrowband filters passing light from hydrogen alpha (H-alpha), oxygen ([OIII]) and sulphur ([SII]), plus red, green and blue broadband filters for star colour. The total exposure time was just over 14 hours.

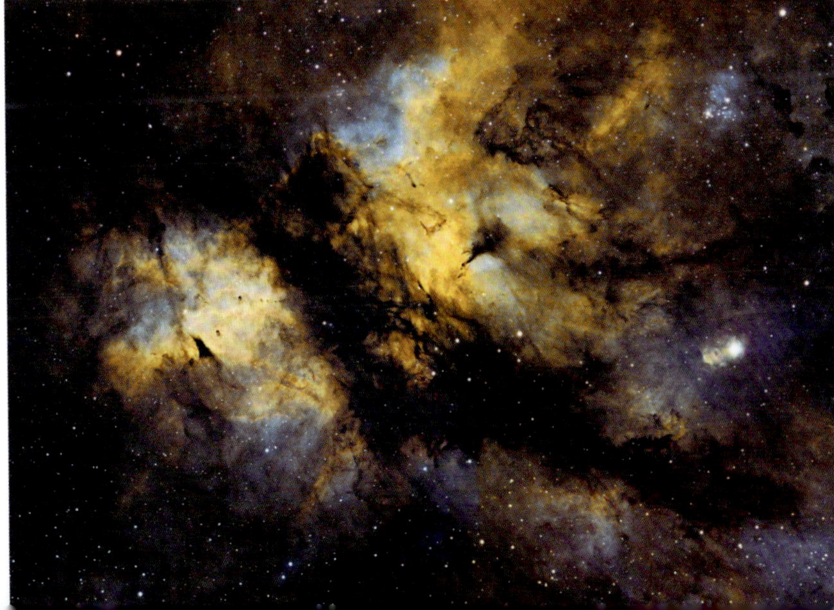

JUNE'S CALENDAR

SUNDAY	MONDAY	TUESDAY	WEDNESDAY	THURSDAY	FRIDAY	SATURDAY
1 Venus W elongation (am); Moon between Regulus and Mars	2	3 4.41 am First Quarter	4	5	6 Moon near Spica	7
8	9 Moon near Antares	10 Major standstill Moon near Antares	11 8.44 am major standstill Full Moon	12	13	14
15	16	17 Mars very near Regulus	18 8.19 pm Last Quarter Moon	19 Moon near Saturn (am)	20	21 Summer Solstice
22 Moon near Venus (am)	23 Moon occults the Pleiades (am)	24	25 11.31 pm New Moon	26	27 Moon near Mercury	28
29 Moon between Mars and Regulus	30					

SPECIAL EVENTS

- **1 June:** Venus reaches its greatest separation from the Sun in the morning sky.
- **1 June:** the crescent Moon lies between Regulus (left) and Mars (right).
- **Night of 10/11 June:** the lowest Full Moon in 19 years, as it reaches a major standstill (see Topic). From London, the Moon reaches an elevation of only 10 degrees in the southern sky. For Shetland, the Full Moon skims little more than a degree above the horizon.
- **17 June:** Mars is less than a degree from Regulus (see Planet Watch and Chart 6a).
- **19 June, before dawn:** the Last Quarter Moon passes just above Saturn.
- **21 June, 3.42 am:** Summer Solstice. The Sun reaches its most northerly point in the sky, so today is Midsummer's Day, with the longest period of daylight and the shortest night.
- **22 June, before dawn:** Venus lies immediately below the crescent Moon (Chart 6b).
- **23 June, 2.30-4.30 am:** the crescent Moon occults some of the stars in the Pleiades, but the event will be hard to observe in the dawn twilight.
- **27 June:** Mercury lies to the right of the crescent Moon.
- **29 June:** the Moon passes between Mars (left) and Regulus (right).

JUNE'S PLANET WATCH

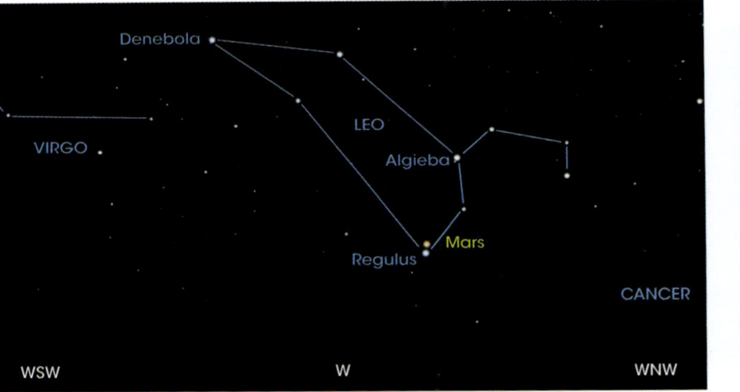

6a *17 June, 11 pm. Mars in conjunction with Regulus.*

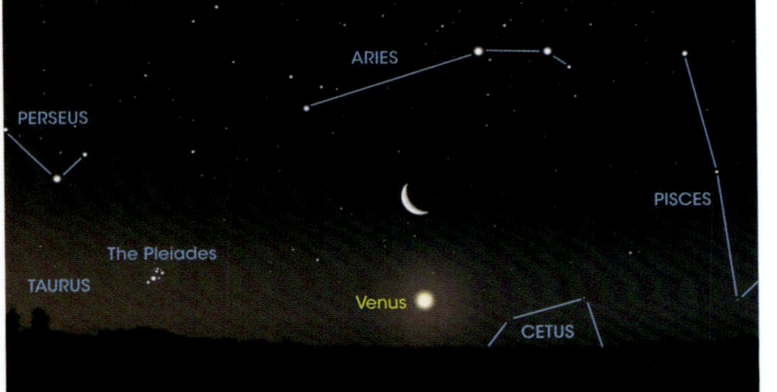

6b *22 June, 3 am. The Moon next to Venus.*

- The only planet easily visible in the evening sky is **Mars**, at magnitude +1.4 and setting around 1 am. The Red Planet is travelling through Leo, and passes only 45 arcminutes above Regulus on 17 June (Chart 6a). The two objects are about the same brightness, but the redness of Mars contrasts strongly with the blue-white star – a lovely sight in binoculars. The Moon lies near Mars on 1 and 29 June.
- From the middle of June, look out for elusive **Mercury** (magnitude –0.5) very low in the north-west after sunset, and setting around 11 pm. On 27 June, you'll find Mercury to the lower right of the thinnest crescent Moon (best seen in binoculars).
- **Saturn**, in Pisces, is rising about 2 am and shines at magnitude +1.0. The Moon is nearby on the morning of 19 June.
- Also rising around 2 am, **Neptune** lies above Saturn, in Pisces. The two worlds are converging, and they appear only a degree apart by the end of the month. The outermost planet glows at a dim magnitude +7.8 and you'll need optical aid to spot it.
- Glorious **Venus**, at a resplendent magnitude –4.1, rises about 3 am. There's a stunning sight on the morning of 22 June, with the crescent Moon hanging right above Venus (Chart 6b).
- **Jupiter** and **Uranus** are lost in the Sun's glare this month.

- The sky at 11 pm in mid-July, with Moon positions at three-day intervals either side of Full Moon.
- The star positions are also correct for midnight at the beginning of July, and 10 pm at the end of the month.
- The planets move slightly relative to the stars during the month.

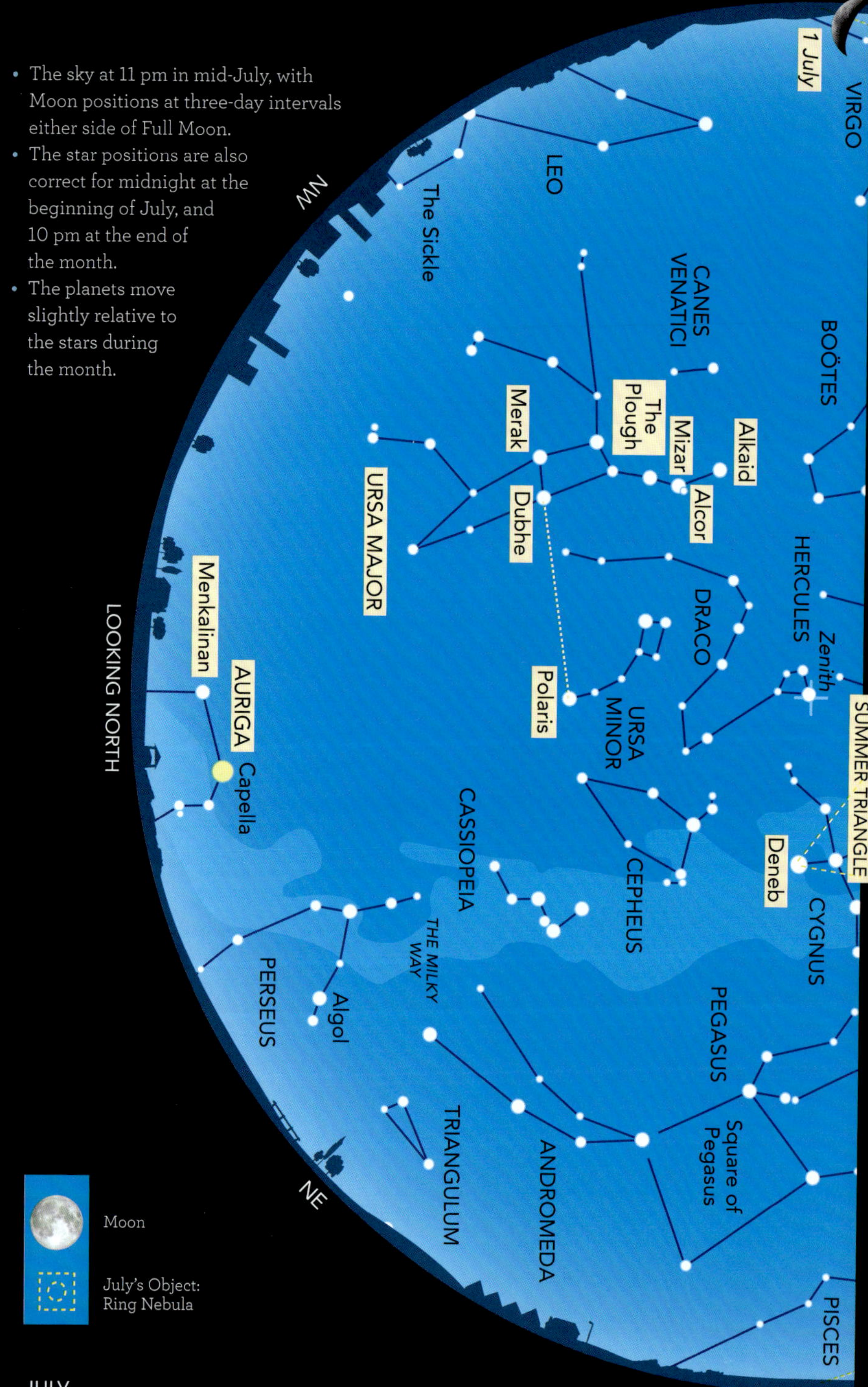

Moon

July's Object: Ring Nebula

JULY

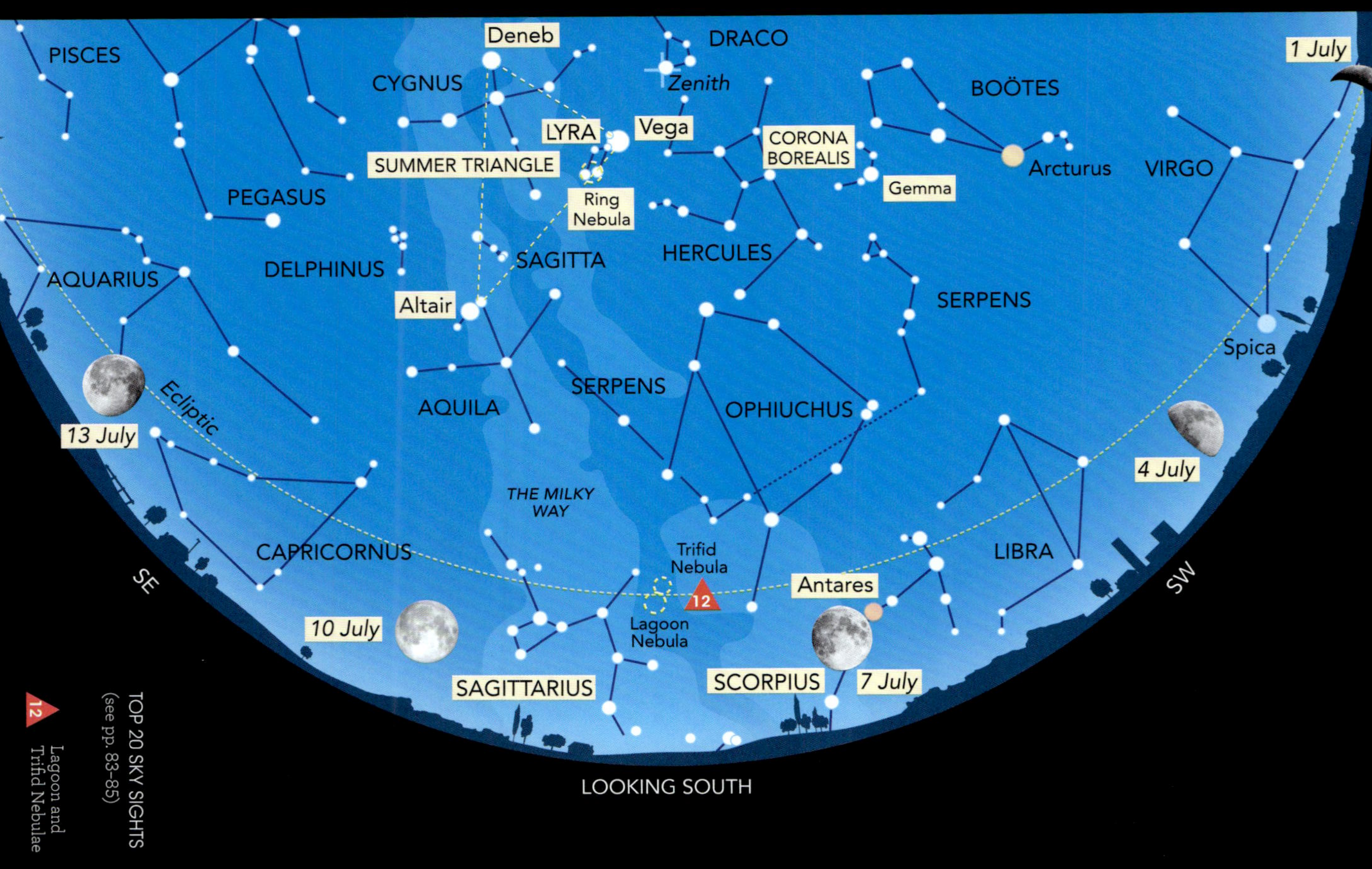

TOP 20 SKY SIGHTS
(see pp. 83–85)

12 Lagoon and Trifid Nebulae

Two of the most ancient constellations are at their best this month, **Sagittarius** and **Scorpius**, decorating the southern region of the sky. Higher up, three brilliant stars – **Vega, Deneb** and **Altair** – mark out the vertices of the **Summer Triangle**.

JULY'S CONSTELLATION

Ursa Major, the Great Bear, is a constellation with deep roots. This star pattern may date back to a period 30,000 years ago when Palaeolithic people in Siberia erected shrines to cave bears. At the end of the last Ice Age, the bear tradition was exported both to southern Europe and to the New World over the Bering Strait.

The seven brightest stars are often called the **Plough** or, in North America, the Big Dipper. You can use its two end stars – **Merak** and **Dubhe** – to locate **Polaris**, the Pole Star or North Star (see Star Chart). In the middle of the bear's tail you'll find one of the few double stars that you can 'split' with your unaided eyes: **Mizar** and its fainter companion **Alcor**. A small telescope reveals Mizar itself to be a close double.

Unlike most constellations, the majority of the stars in the Plough lie at the same distance and were born together. Leaving aside Dubhe and **Alkaid**, the others are all moving in the same direction, along with other stars of the Ursa Major Moving Group, which include **Menkalinan** in **Auriga** and **Gemma** in **Corona Borealis**. Over thousands of years, the shape of the Plough will gradually change, as Dubhe and Alkaid go off on their own paths.

OBSERVING TIP

This is the month when you really need a good, unobstructed horizon to the south, for the best views of the glorious summer constellations of Scorpius and Sagittarius. They never rise high in temperate latitudes, so make the most of a southerly view – especially over the sea – if you're away on holiday. A good southern horizon is also best for views of the planets, because they rise highest when they're in the south.

JULY'S OBJECT

Looking like a celestial smoke ring, the **Ring Nebula** lies between the two lowest stars of the tiny constellation **Lyra**, below brilliant **Vega**. You can spot it in a small telescope, but the Ring Nebula is only 1 arcminute across and a 200-mm instrument is needed to see its fine details.

The Ring Nebula is the wraith of a star that died thousands of years ago, when its core ran out of nuclear fuel, and its outer layers puffed off into space. The James Webb Space Telescope has revealed that it contains 20,000 dense globules of gas, and is filled with carbon-rich compounds forged in the old star. In future years, these molecules may form part of a new planetary system, where they could give rise to life.

JULY'S TOPIC: SATELLITE CONSTELLATIONS

Showing my friends the odd satellite crossing the sky used to be one of the fun aspects of star parties, especially if the spacecraft was prone to sudden flare-ups or – even better – carrying astronauts. But now satellites just fill my heart with dread.

The problem? Communications companies are starting to launch satellites by the thousands to improve mobile phone and internet services to even the most remote regions of the world. In itself, that is an excellent idea, and most astronomers don't want to impede it.

But imagine a sky where you can't spot the constellation patterns for hundreds of moving star-like objects in front. Where you can't take a picture – like those featured in this book – because it will be covered in long streaks from passing satellites. And where professional astronomers won't be able to probe the Universe – or discover asteroids about to hit the Earth – because bright satellites are photobombing their deep exposures.

According to the International Astronomical Union Centre for the Protection of the Dark and Quiet Sky from Satellite Constellation Interference (mercifully abbreviated to CPS), as of early 2024 there were 6105 of these satellites in orbit. But the planned total – so far – is 537,267 satellites! You can keep track of the running totals at cps.iau.org.

Fortunately, the satellite operators are listening to astronomers. The largest, SpaceX, is deliberately painting its satellites black, to keep them below naked-eye visibility, and also orientating their solar panels so they don't reflect dazzling sunlight down to Earth when it's night-time below.

On 2 March 2023, Jo Bourne captured the stunning conjunction of Venus and Jupiter over Stonehenge, with a Sony Alpha 7 III camera and FE 24-105-mm f/4 lens at 60-mm. It was an 8-second exposure at f/8 and ISO 100. She used a Starglow filter for 2 seconds of the exposure to make the planets stand out.

JULY'S PICTURE

A conjunction of the planets is always an impressive sight – and truly spectacular when we witness the two most brilliant worlds appear to kiss. The brighter planet here is Venus, flying past Jupiter early in 2023. Despite appearances, there was no danger of a cosmic collision as the giant planet lay four times further away than the Evening Star.

Stonehenge provides an impressive foreground for Jo Bourne's evocative image, though any association of this ancient monument in Wiltshire with the planets has been lost over the millennia. We do know from surviving documents, though, that Mayan stone structures in Mexico were used to observe Venus.

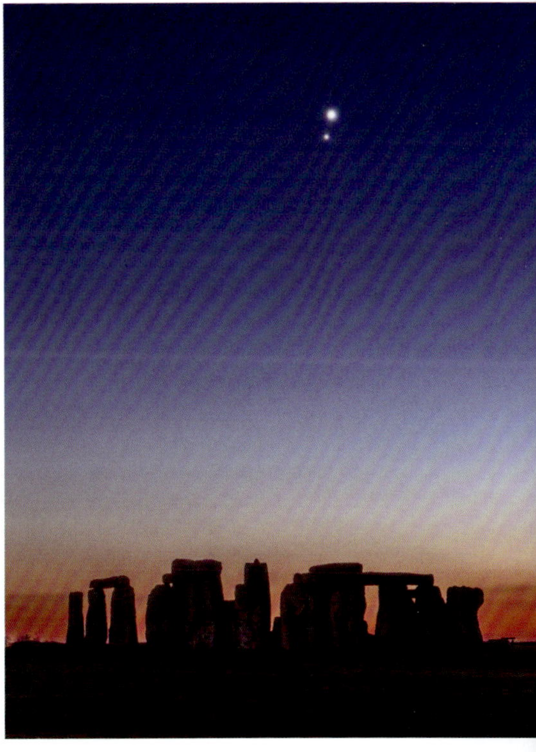

JULY'S CALENDAR

SUNDAY	MONDAY	TUESDAY	WEDNESDAY	THURSDAY	FRIDAY	SATURDAY
		1	2 8.30 pm First Quarter Moon	3 Earth at aphelion; Moon near Spica	4 Venus, Uranus, Pleiades in line (am); Mercury E elongation	5
6 Saturn closest to Neptune	7 Moon near Antares	8	9	10 9.37 pm Full Moon	11	12
13	14	15 Moon near Saturn	16	17	18 1.38 am Last Quarter Moon	19
20	21 Moon near Venus (am)	22 Moon near Venus (am)	23 Moon near Jupiter (am)	24 8.11 pm New Moon	25	26
27	28 Moon near Mars	29	30 Moon near Spica	31		

SPECIAL EVENTS

- **3 July, 8.55 pm:** the Earth is furthest from the Sun (aphelion), at 152 million km.
- **4 July, before dawn:** the Pleiades, Uranus and Venus are in line, providing an ideal chance to spot the seventh planet (see Planet Watch and Chart 7a).
- **6 July:** Saturn approaches within 1 degree of Neptune (see Planet Watch).
- **15 July:** the Moon lies to the right of Saturn.
- **21 July, before dawn:** you'll find brilliant Venus below the crescent Moon, with Aldebaran to the right (Chart 7b).
- **22 July before dawn:** the Moon is flanked by the two brightest planets, Venus to the right and Jupiter to the lower left (Chart 7b).
- **23 July before dawn:** the bright 'star' below the Moon is Jupiter, with Venus well to the right (Chart 7b).
- **28 July:** Mars lies above the crescent Moon.

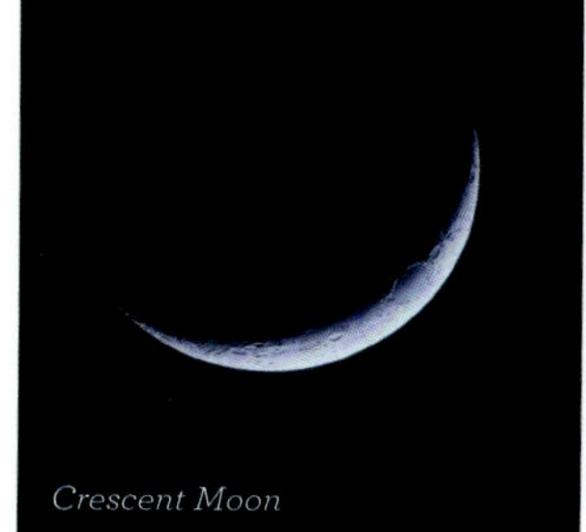
Crescent Moon

JULY'S PLANET WATCH

7a 4 July, 3 am. Uranus between Venus and the Pleiades.

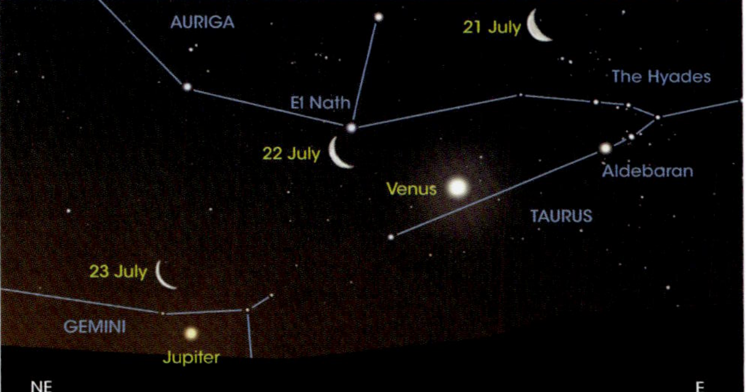

7b 21–23 July, 3.45 am. The crescent Moon passes Venus and Jupiter.

- You'll find **Mars** low in the west after sunset in Leo, setting around 11 pm. Currently shining at magnitude +1.5, the Red Planet is gradually fading as the Earth pulls away from it. The crescent Moon passes just below Mars on 28 July.
- Rising about 11.30 pm, **Saturn** lies in Pisces and shines at +0.9. The Moon is nearby on 15 July.
- **Neptune** (magnitude +7.7) lies only a degree from Saturn all month; it's also in Pisces and rises around 11.30 pm. The two worlds are closest – just 58 arcminutes apart – on 6 July. You can find the outermost planet in binoculars or a small telescope, as a faint bluish 'star' two Moon-widths above Saturn.
- **Uranus** is in Taurus, rising around 1.30 am. The seventh planet lies just below the Pleiades all month, and on the morning of 4 July it's directly on a line connecting Venus and the Pleiades – an ideal opportunity to locate the dim world (magnitude +5.8) using binoculars or a small telescope (Chart 7a).
- The brilliant Morning Star **Venus** rises about 2 am, at magnitude –4.1. The crescent Moon forms a lovely pairing with Venus on the mornings of 21 and 22 July (Chart 7b).
- This month, the giant planet **Jupiter** emerges from the glare of the Sun into the morning sky. During the second half of July, it appears low in the north-eastern sky about 3.30 am, well to the lower left of Venus. Brighter than any star, at magnitude –1.9, Jupiter lies in Gemini. A slender crescent Moon is nearby on the morning of 23 July (Chart 7b) – best seen in binoculars.
- **Mercury** is hidden in the bright twilight glow, even though it reaches its greatest separation from the Sun on 4 July.

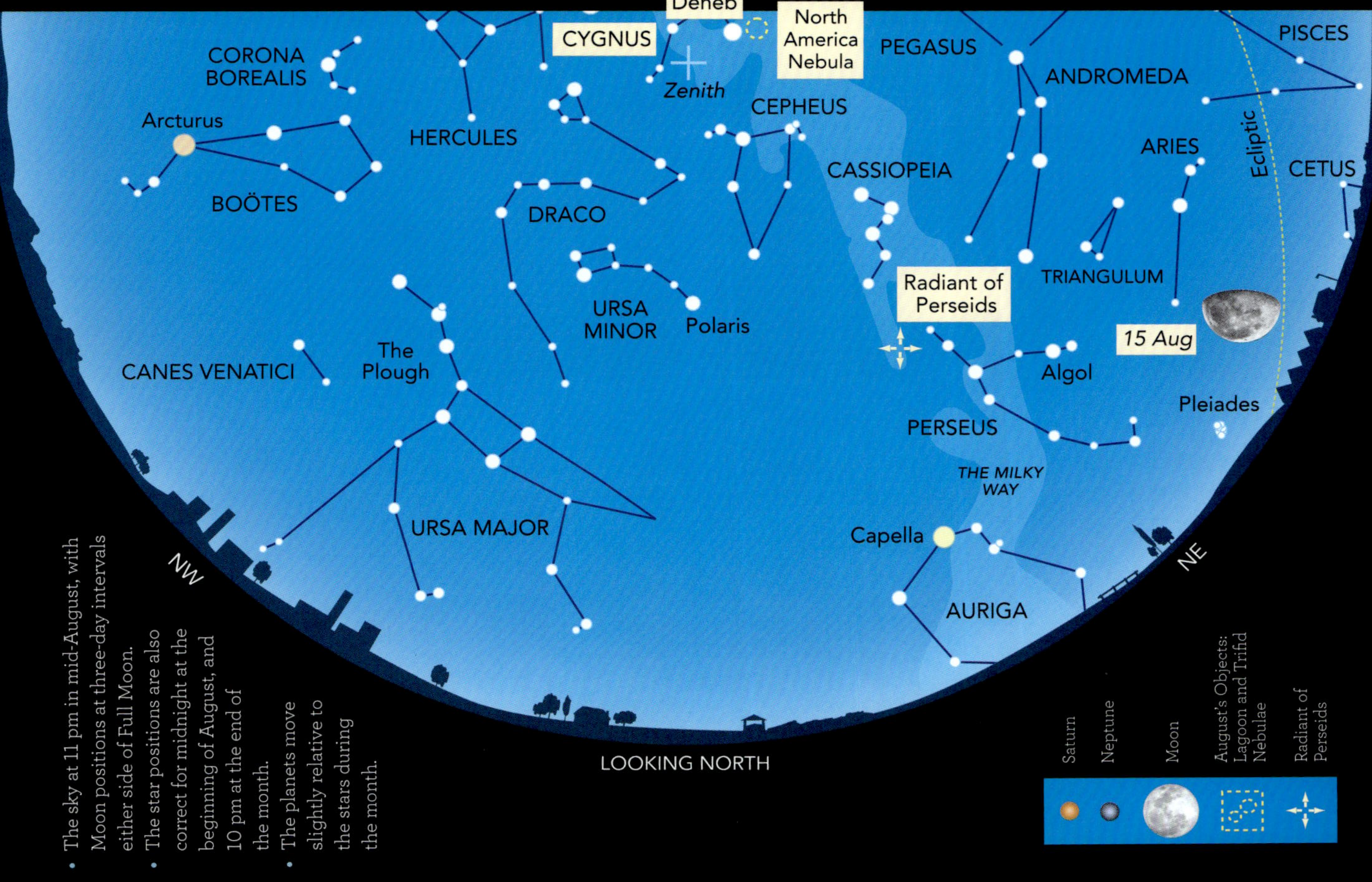

- The sky at 11 pm in mid-August, with Moon positions at three-day intervals either side of Full Moon.
- The star positions are also correct for midnight at the beginning of August, and 10 pm at the end of the month.
- The planets move slightly relative to the stars during the month.

EAST
WEST
LOOKING SOUTH
SE
SW
ANDROMEDA
DRACO
CORONA BOREALIS
PISCES
North America Nebula
Zenith
Deneb
Vega
Square of Pegasus
Butterfly Nebula
LYRA
12 Aug
CYGNUS
Albireo
HERCULES
SERPENS
Dumbbell Nebula
Great Rift
PEGASUS
SAGITTA
LIBRA
Neptune
Saturn
DELPHINUS
Altair
OPHIUCHUS
CETUS
AQUARIUS
AQUILA
SERPENS
SCORPIUS
THE MILKY WAY
9 Aug
Trifid Nebula
Ecliptic
PISCIS AUSTRINUS
Lagoon Nebula
CAPRICORNUS
6 Aug
SAGITTARIUS
13
14
TOP 20 SKY SIGHTS
(see pp. 83–85)
13 Albireo
14 Dumbbell Nebula

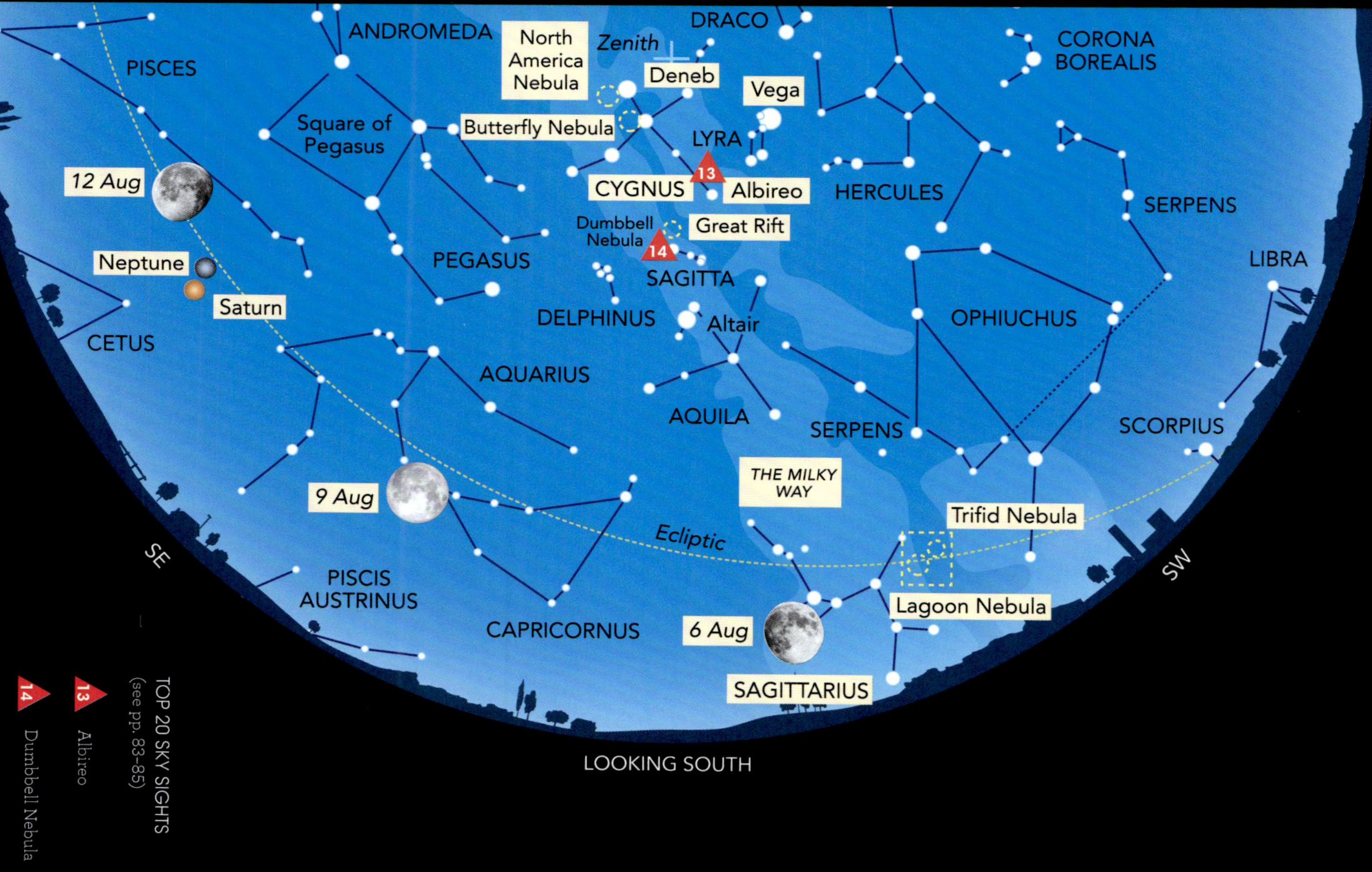

AUGUST

There's a 'Glorious Twelfth' this August, when Venus joins Jupiter for a spectacular conjunction of the two most brilliant planets in the morning sky. Later in the month, the show continues when they are joined by Mercury and the crescent Moon.

AUGUST'S CONSTELLATION

The flying swan – **Cygnus** – actually looks like its namesake, with outspread wings and an elongated neck. According to myth, the great god Zeus disguised himself as a swan to seduce Leda, the wife of King Tyndareus of Sparta; her mixed mortal and immortal offspring included Castor and Pollux (commemorated in the stars of Gemini) and the legendary beauty Helen of Troy.

Deneb marks the swan's tail. It's the most distant of the first-magnitude stars, lying 1500 light years from the Earth and shining with the luminosity of 55,000 Suns. If it lay as close to us as its neighbour **Vega**, Deneb would appear 15 times brighter than Venus.

The swan's head is marked by **Albireo**, to my mind the most beautiful double star in the sky. The **Milky Way** meanders through Cygnus, split by the dark silhouette of the **Great Rift**. On a dark night, sweep along the constellation with binoculars or a small telescope to reveal a treasure trove of star clusters and glowing gas clouds, including the **North America Nebula** and the **Butterfly Nebula** (see June's Picture).

AUGUST'S OBJECTS

A pair of glorious nebulae adorn the constellation **Sagittarius**. The brighter is the **Lagoon Nebula**, just visible to the unaided eye as a glowing cloud three times the width of the Moon. Lying about 5000 light years away, the nebula is illuminated by a newly born star that's 40 times heavier than the Sun and 200,000 times brighter. Dark islands in the glowing lagoon are dense dust clouds, each poised to condense into new stars and planets.

Its smaller companion, the **Trifid Nebula**, looks dramatically different. It's split into three lobes (the meaning of the word 'trifid') by bands of dark cosmic dust silhouetted against its bright core. The nebula is packed with young stars and embryonic stars that have yet to fully form.

OBSERVING TIP

It's always fun to search out 'faint fuzzies' in the sky – star clusters, nebulae and galaxies. But don't even think of observing these objects around the time of Full Moon, as its light will drown them out. You'll have the best views near New Moon: a period astronomers call 'dark of Moon'. When the Moon is bright, focus on planets, bright double stars – and, of course, the Moon itself.

AUGUST'S TOPIC: THE CELESTIAL POLICE

In the late eighteenth century, astronomers began to suspect there was a missing planet in the Solar System. The distances from the Sun to the known planets – including recently discovered Uranus – formed a regular progression, except for a gap between the orbits of Mars and Jupiter.

After spending a fruitless decade searching for this faint world, Baron Franz Xaver von Zach in Hungary decided to call in extra forces in a systematic crackdown. He divided the sky into 24 regions, and recruited 23 other astronomers in 'a strictly organised policing of the heavens', each repeatedly patrolling their own patch 'to confirm the unchanging state of his district, or the presence of each wandering foreign guest'.

By the time the invitation to join the self-styled Celestial Police reached Giuseppe Piazzi at the Palermo Observatory in Sicily, however, Piazzi had already made a citizen's arrest. On 1 January 1801 he had noted a 'star' in Taurus that wasn't in any existing catalogue. By the following night it had moved, and Piazzi concluded it had to be a faint world in the Solar System.

Piazzi named this 'minor planet' Ceres, for the patron goddess of Sicily. The Celestial Police carried on to apprehend another three minor planets – now more commonly called 'asteroids' – orbiting between Mars and Jupiter. Today, the number of known asteroids has grown to over a million, of which 23,000 have been named.

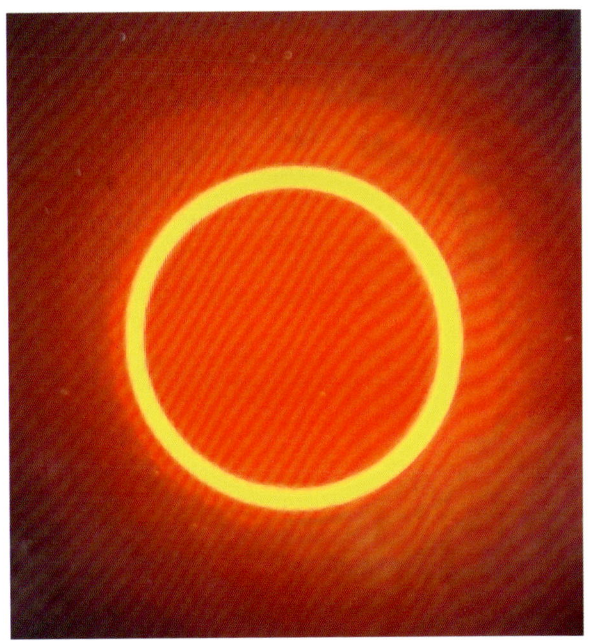

With a Nikon D7200 camera and Nikkor 18-140-mm lens at f/5.6 and ISO 100, and an exposure of 0.1 second through a Celestron EclipSmart solar filter, I photographed the midpoint of the annular eclipse on 14 October 2023. To share the view in real time, I then imaged the rear screen of the camera with a Samsung Galaxy A53 phone (on automatic settings) – unorthodox processing that enhanced the fieriness of the solar ring.

AUGUST'S PICTURE

A stunning 'Ring of Fire' appeared in the morning sky over parts of North America in October 2023, as the Sun suffered an annular eclipse – and I travelled to Piute State Park in Utah to witness the solar spectacle.

During a total solar eclipse, the Moon appears the same size as the Sun in the sky, and as it moves in front the Moon totally blots out the Sun's brilliant face to reveal its faint outer atmosphere, the corona. But on this occasion, the Moon was at the far point of its orbit and appeared smaller than the Sun. As a result, a ring – or annulus – of the Sun's bright surface appeared around the Moon's silhouette. Annular eclipses are just as common as total eclipses and – while not as spectacular – a Ring of Fire eclipse has its own magic.

AUGUST'S CALENDAR

SUNDAY	MONDAY	TUESDAY	WEDNESDAY	THURSDAY	FRIDAY	SATURDAY
31 7.25 am First Quarter Moon near Antares					1 1.41 pm First Quarter Moon	2
3 Moon near Antares	4	5	6	7	8	9 8.55 am Full Moon
10	11 Moon near Saturn	12 Venus very near Jupiter (am); Moon near Saturn; Perseids	13 Perseids (am)	14	15	16 6.12 am Last Quarter Moon
17 Moon near Pleiades and Aldebaran (am)	18	19 Mercury W elongation	20 Moon between Venus and Jupiter	21 Moon near Mercury, Venus and Jupiter	22	23 7.06 am New Moon
24	25	26	27	28	29	30

SPECIAL EVENTS

- **11 August:** the Moon lies to the right of Saturn.
- **12 August, before dawn:** a close conjunction of the two brightest planets, as Venus passes less than a degree from Jupiter (see Planet Watch and Chart 8a).
- **12 August:** the Moon is to Saturn's left.
- **Night of 12/13 August:** maximum of the **Perseid meteor shower**. Sadly, moonlight washes out this prolific display of shooting stars, high-speed pellets of dust from Comet Swift–Tuttle burning up high above our heads.
- **17 August, early hours:** the Moon passes above Aldebaran, with the Pleiades to its right.
- **19 August:** Mercury reaches its greatest separation from the Sun (see Planet Watch).
- **20 August, before dawn:** the crescent Moon lies between the two brightest planets, Venus and Jupiter (see Planet Watch and Chart 8b).
- **21 August, before dawn:** the slim crescent Moon passes above Mercury, to the left of Venus and Jupiter (see Planet Watch and Chart 8b).

AUGUST'S PLANET WATCH

8a 12 August, 4 am. Close conjunction of Venus and Jupiter (binocular view).

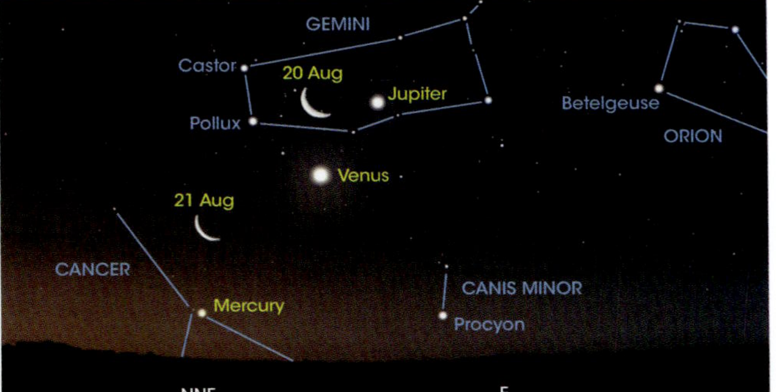

8b 20–21 August, 4.45 am. The Moon passes Jupiter, Venus and Mercury.

- **Mars** lies in Virgo, shining at magnitude +1.5 and setting about 10.30 pm. During August, the Red Planet slips down into the twilight glow and won't be visible again until next year.
- At magnitude +0.7, **Saturn** – in Pisces – is rising around 9.30 pm. The Moon is nearby on 11 and 12 August.
- **Neptune** (magnitude +7.7) is also in Pisces and rising about 9.30 pm.
- **Uranus** glows at a slightly brighter magnitude +5.8 in Taurus, 4 degrees below the Pleiades. The seventh planet rises around 11.30 pm.
- Before dawn, we're treated to the wonderful sight of the two most brilliant planets rising together: look east between 3 and 4 am. During the first few days of August, glorious Venus (magnitude –4.0) is the higher of the two, with **Jupiter** seven times fainter at magnitude –1.9 – but still outshining any star in the sky.
- On the morning of 12 August, **Venus** glides just 52 arcminutes below Jupiter. This close conjunction will be magnificent as seen in binoculars or a low-power telescope (Chart 8a). After that Jupiter rises higher in sky, among the stars of Gemini, while Venus gradually sinks towards the horizon.
- Look out for a pretty sight on the morning of 20 August, when the crescent Moon lies between Jupiter and Venus. There's another predawn spectacle on 21 August, with the slender Moon above Mercury (see below), with Venus and Jupiter to the right (Chart 8b).
- Late in August, **Mercury** appears in the morning sky to the lower left of Venus. At its greatest separation from the Sun on 19 August, Mercury rises at 4.15 am, shining at magnitude 0.0; it brightens to magnitude –1.2 by the end of August.

- The sky at 11 pm in mid-September, with Moon positions at three-day intervals either side of Full Moon.
- The star positions are also correct for midnight at the beginning of September, and 10 pm at the end of the month.
- The planets move slightly relative to the stars during the month.

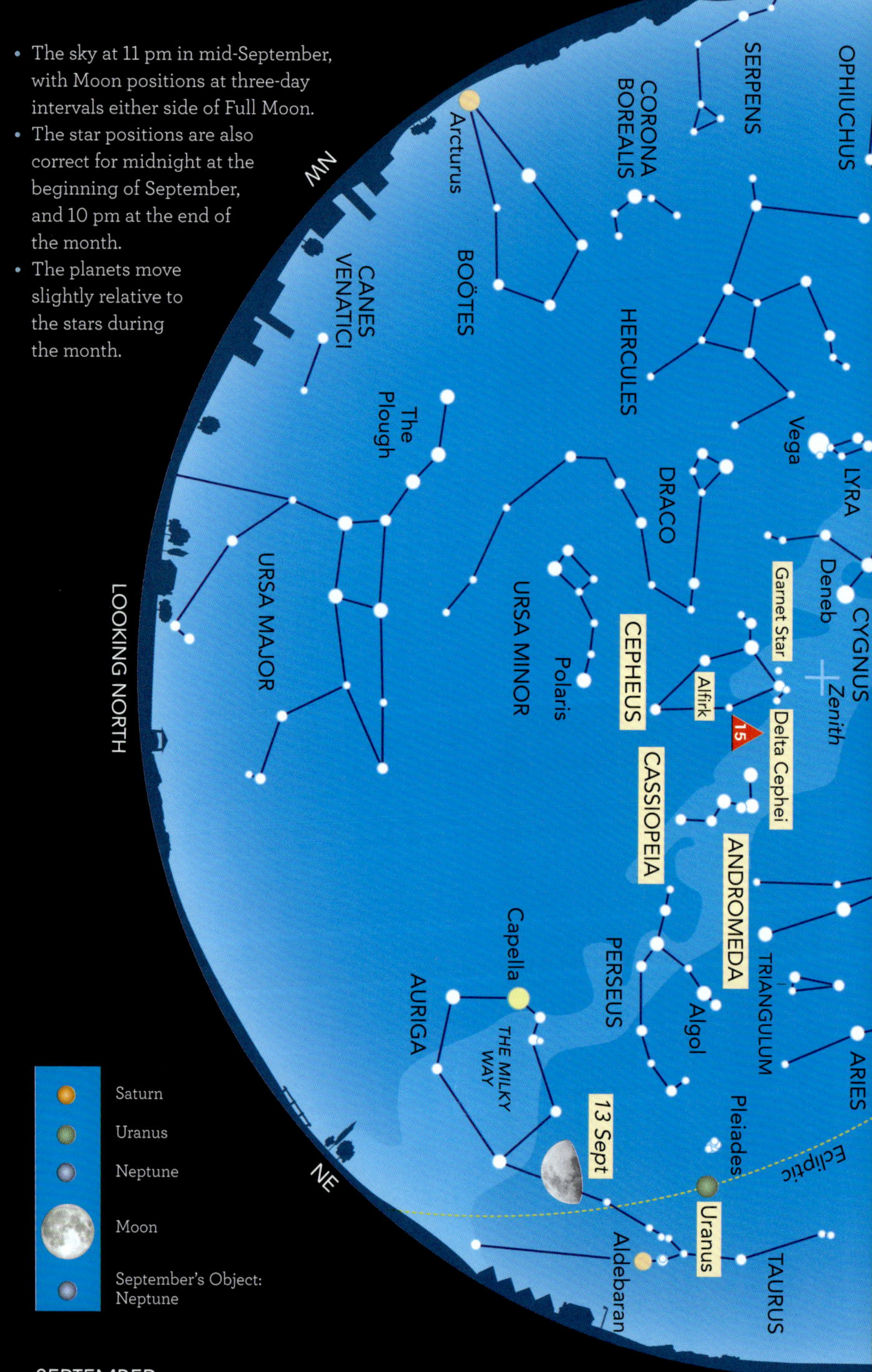

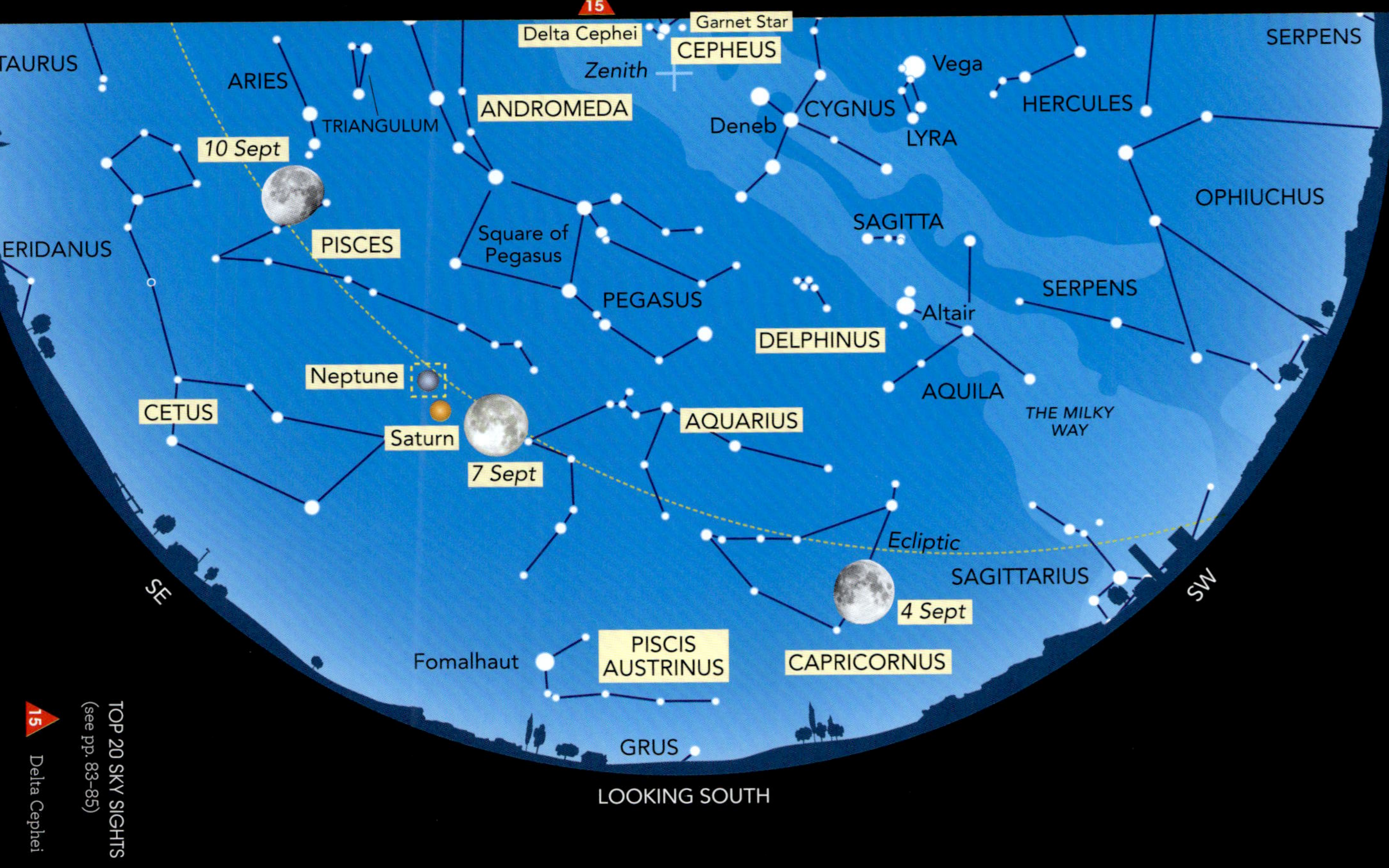

TOP 20 SKY SIGHTS
(see pp. 83–85)

15 Delta Cephei

SEPTEMBER

The Moon grabs the headlines this month, with a total lunar eclipse and an occultation of the Seven Sisters. **Saturn** is also at its nearest and brightest. On the starry front, the sky is awash with watery constellations: **Piscis Austrinus** (the Southern Fish) and **Aquarius** (the Water Carrier) are joined by **Delphinus** (the Dolphin), the strangely named Sea Goat (**Capricornus**), a pair of Fishes (**Pisces**) and **Cetus** (the Sea Monster).

SEPTEMBER'S CONSTELLATION

Shaped like a child's drawing of a house, **Cepheus** represents a mythical king of Ethiopia. He is overshadowed – both in legend and as a constellation – by his wife, Queen **Cassiopeia**, a bright star pattern representing the proud mother of Princess **Andromeda**.

Three fascinating stars lift Cepheus from near obscurity. **Alfirk** is a double star, its companion visible through a small telescope. The aptly named **Garnet Star** is a red hypergiant, some 1000 times wider than the Sun, which varies slightly in brightness over the years.

The star of the show, though, is **Delta Cephei**. As this giant star pulsates in size, its brightness changes from magnitude +3.5 to +4.4 over a period of five days and nine hours. It's the prototype of the 'cepheid variable stars', whose period of variation is closely related to their intrinsic brilliance. As a result, astronomers can use these stellar beacons to measure distances way beyond our Galaxy.

SEPTEMBER'S OBJECT

With Pluto being demoted to a mere 'ice dwarf', **Neptune** is officially the most remote planet in our Solar System. It lies 4500 million kilometres out – 30 times the Earth's distance – in the twilight zone of our family of worlds, and takes nearly 165 years to circle the Sun.

OBSERVING TIP

This is the ideal time of year to tie down the main compass directions, as seen from your observing site. North is easy – just latch onto Polaris, the Pole Star, using the familiar stars of the Plough (see July's Constellation). And at noon, the Sun is always towards the south. But the useful extra in September is that we hit the Autumn Equinox, when the Sun rises due east, and sets due west. Memorise those positions relative to a tree or house around your horizon, and you can orientate yourself correctly once night has fallen.

Neptune is at its closest this year on 23 September, and visible in Pisces through a small telescope. But you need a spacecraft to unveil the secrets of this watery world. In 1989, Voyager 2 sent us close-up views of a turquoise planet 17 times heavier than Earth, adorned by clouds of methane and ammonia, and encircled by faint rings of dusty debris. And the spacecraft returned stunning views of Neptune's moons, including icy volcanoes on its giant moon Triton.

For a world so far from the Sun, Neptune is amazingly frisky. Its core blazes at over 5000°C – almost as hot as the Sun's surface. This internal heat drives dramatic storms and the fastest winds in the Solar System, raging at 2100 kilometres per hour.

SEPTEMBER'S TOPIC: THE MOON'S CRATERS

Even through a modest pair of binoculars, you can see that the Moon is covered in craters. First to describe them was Galileo, in 1609. Some 40 years later, fellow Italian astronomer Giovanni Battista Riccioli gave names to 247 craters, immortalising eminent scientists and philosophers – including himself!

Until the 1950s, many astronomers believed the lunar craters were extinct volcanoes. After robotic spacecraft and the Apollo astronauts visited the Moon, scientists realised the craters were blasted out by asteroids and comets that brutally bombarded the Moon in its infancy.

The Earth suffered, too. But weather and geological activity have smoothed over our planet's cosmic scars. Not so the Moon. With virtually no atmosphere, its impact history is there for all to see. The most ferocious assault took place 3.8 billion years ago, during the Late Heavy Bombardment. This gouged out enormous craters, up to 1200 kilometres across, later filled by dark lava to make up the face of the 'Man in the Moon'.

Best time to see craters? When the Moon is half-lit, and the Sun's illumination comes side-on.

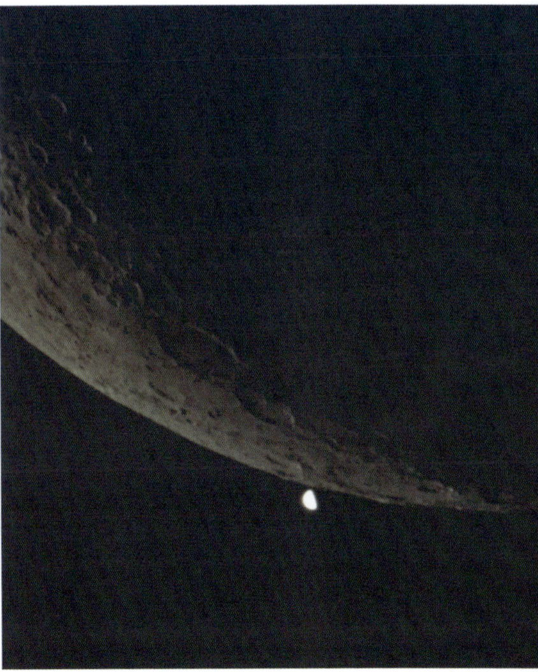

Dave Eagle captured the Moon encroaching on Venus just before an occultation during daylight on the morning of 9 November 2023, using a Sky-Watcher Esprit 120ED 120-mm refractor, with a 2× Barlow lens and a ZWO ASI585MC camera. This is one of a series of images that he produced by videoing the full event, and then running the video through AutoStakkert! software which automatically selects and stacks the sharpest frames.

SEPTEMBER'S PICTURE

The headlines always proclaim when the Moon moves in front of the Sun in a solar eclipse. Less widely touted are the occasions when the Moon hides another planet in our Solar System; but these 'planetary occultations' are thrilling in their own way.

Even with the naked eye it's exciting to view a brilliant world like Mars or Jupiter fade from view as it slips behind the lunar limb. And a telescope highlights the contrast between our nearby, cratered and rather unreflective Moon and these distant exotic worlds.

Going back to November 2023, Dave Eagle's image of a lunar occultation of Venus shows just how drab our Moon appears alongside the brilliance of the cloud-covered planet. Venus is in reality four times wider than the Moon; it appears minuscule in this view only because it was 300 times further from us.

SEPTEMBER'S CALENDAR

SUNDAY	MONDAY	TUESDAY	WEDNESDAY	THURSDAY	FRIDAY	SATURDAY
	1 Venus near Praesepe (am)	2	3	4	5	6
7 7.09 pm Full Moon; total lunar eclipse	8 Moon near Saturn and Neptune	9	10	11	12 Moon occults the Pleiades	13
14 11.33 am Last Quarter Moon	15	16 Moon near Jupiter (am)	17 Moon near Jupiter (am)	18	19 Moon near Venus and Regulus (am)	20
21 8.54 pm New Moon; partial solar eclipse; Saturn opposition	22 Autumn Equinox; Neptune closest to Earth	23 Neptune opposition	24	25	26	27
28	29	30 0.54 am First Quarter Moon				

SPECIAL EVENTS

- **1 September, before dawn:** Venus passes just below Praesepe (see Planet Watch).
- **7 September:** a total lunar eclipse is visible from Australia, Asia, Africa and Europe. From the UK and Ireland, the Moon rises just as totality is ending (7.53 pm), with the partial phase ending at 8.56 pm).
- **8 September:** the Moon lies near Saturn and Neptune (see Planet Watch).
- **12 September, 9 pm–12 midnight:** the Moon moves right in front of the Pleiades star cluster (Chart 9a).
- **16 September, early hours:** Jupiter lies below the Moon, with Castor and Pollux to the left (Chart 9b).
- **17 September, early hours:** the crescent Moon is near Jupiter, Castor and Pollux (Chart 9b).
- **19 September, before dawn:** Venus lies only 48 arcminutes from Regulus, with the crescent Moon nearby.
- **21 September:** Saturn is opposite to the Sun, and nearest to the Earth at 1279 million km (see Planet Watch).
- **21 September:** a partial solar eclipse (85%) is visible from the southern Pacific Ocean and Antarctica. Nothing is visible from the UK or Ireland.
- **22 September, 7.19 pm:** the Autumn Equinox.
- **22 September:** Neptune at its closest – 4321 million km (see Planet Watch).
- **23 September:** Neptune lies opposite the Sun in the sky (see Planet Watch).

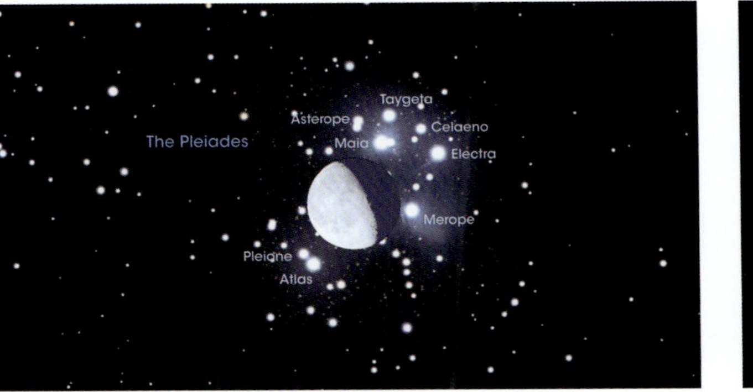

9a 12 September, 10.30 pm. The Moon occults the Pleiades.

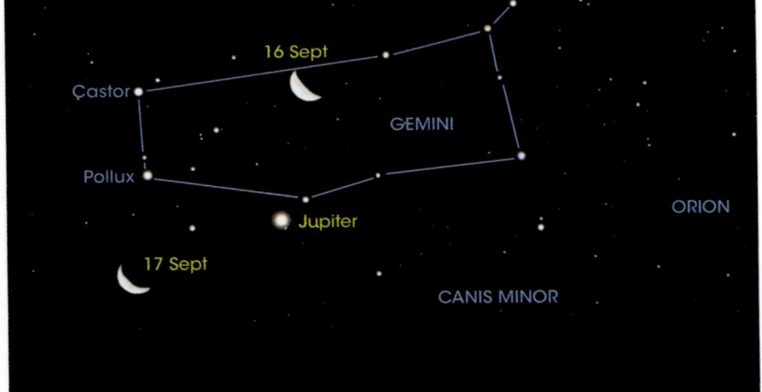

9b 16–17 September, 2 am. The Moon passes near Jupiter, in Gemini.

- **Saturn** is closest to the Earth and opposite the Sun on 21 September (see Special Events), when the planet reaches its maximum brightness this year, at magnitude +0.6. It's visible all night long in Pisces. A small telescope will show Saturn's famous rings – currently appearing almost edge-on – along with its biggest moon, Titan. The Moon passes above Saturn on 8 September.
- Some 2.5 degrees to the left of Saturn, and also in Pisces, **Neptune** is closest to the Earth on 22 September and reaches opposition on 23 September (see Special Events), It's above the horizon all night. Even at its nearest, though, this distant world is below naked-eye visibility (magnitude +7.7) so you'll need binoculars or a telescope to see it. On 8 September, Neptune lies between Saturn and the Moon.
- Its near twin, **Uranus**, lies in Taurus, near the Pleiades, and is a little brighter at magnitude +5.7. The seventh planet rises about 9.30 pm.
- At magnitude –2.0, **Jupiter** is a brilliant jewel adorning Gemini: you'll find the constellation's principal stars, Castor and Pollux, to the upper left of the giant planet. Jupiter rises about 0.30 am, and the Moon is nearby on the mornings of 16 and 17 September (Chart 9b).
- Glorious **Venus** – magnitude –3.9 – rises around 4 am. The Morning Star starts the month in Cancer, and on the morning of 1 September, it's just a degree away from Praesepe, the Beehive star cluster – a lovely sight in binoculars or a small telescope. Travelling into Leo, Venus passes only 45 arcminutes from Regulus on the morning of 19 September, with the crescent Moon nearby.
- **Mercury** and **Mars** appear too close to the Sun to be visible in September.

SEPTEMBER'S PLANET WATCH

WEST

- The sky at 11 pm in mid-October, with Moon positions at three-day intervals either side of Full Moon.
- The star positions are also correct for midnight at the beginning of October, and 9 pm at the end of the month (after the end of BST and IST).
- The planets move slightly relative to the stars during the month.

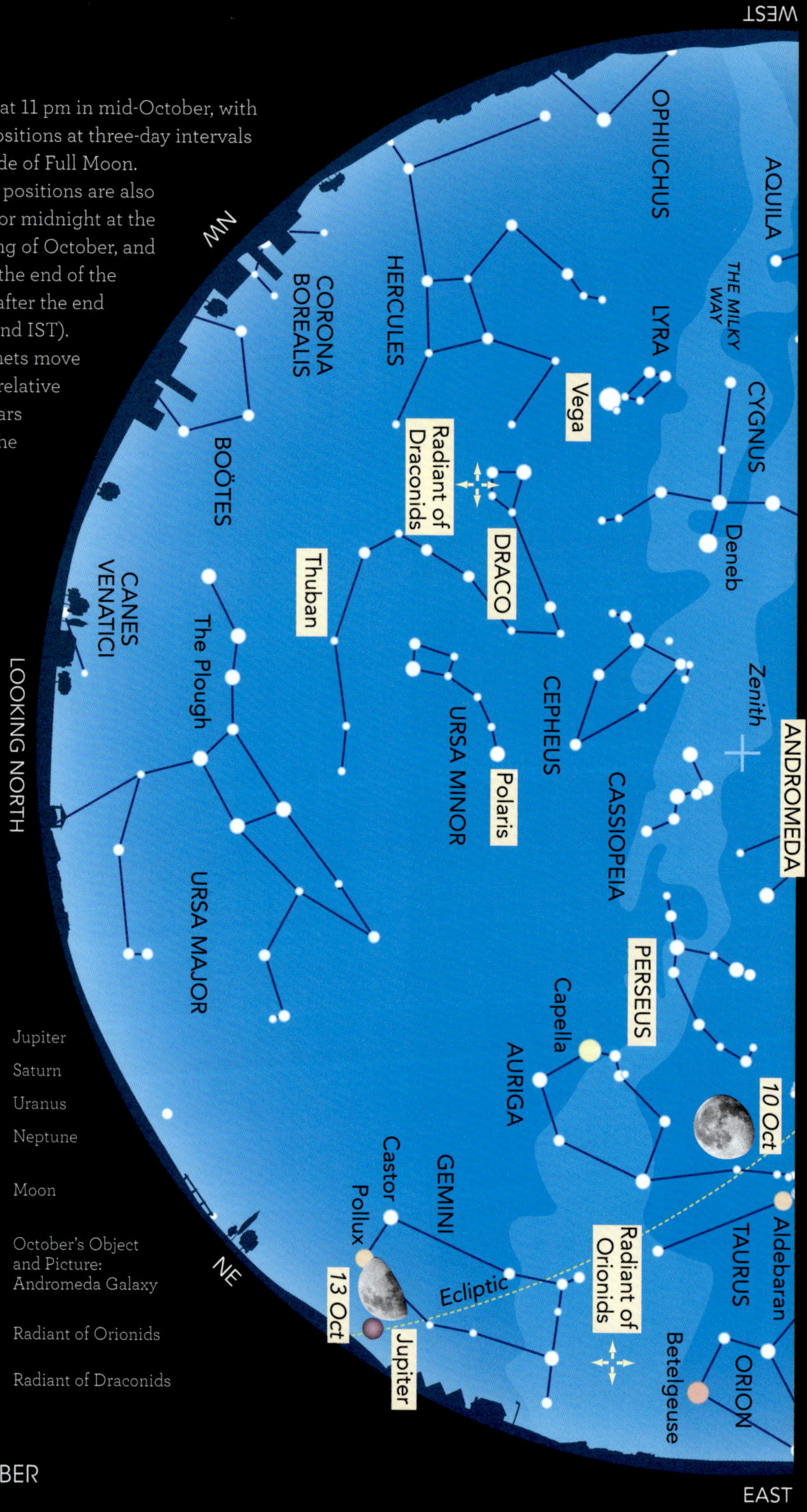

LOOKING NORTH

NW

NE

Jupiter

Saturn

Uranus

Neptune

Moon

October's Object and Picture: Andromeda Galaxy

Radiant of Orionids

Radiant of Draconids

EAST

OCTOBER

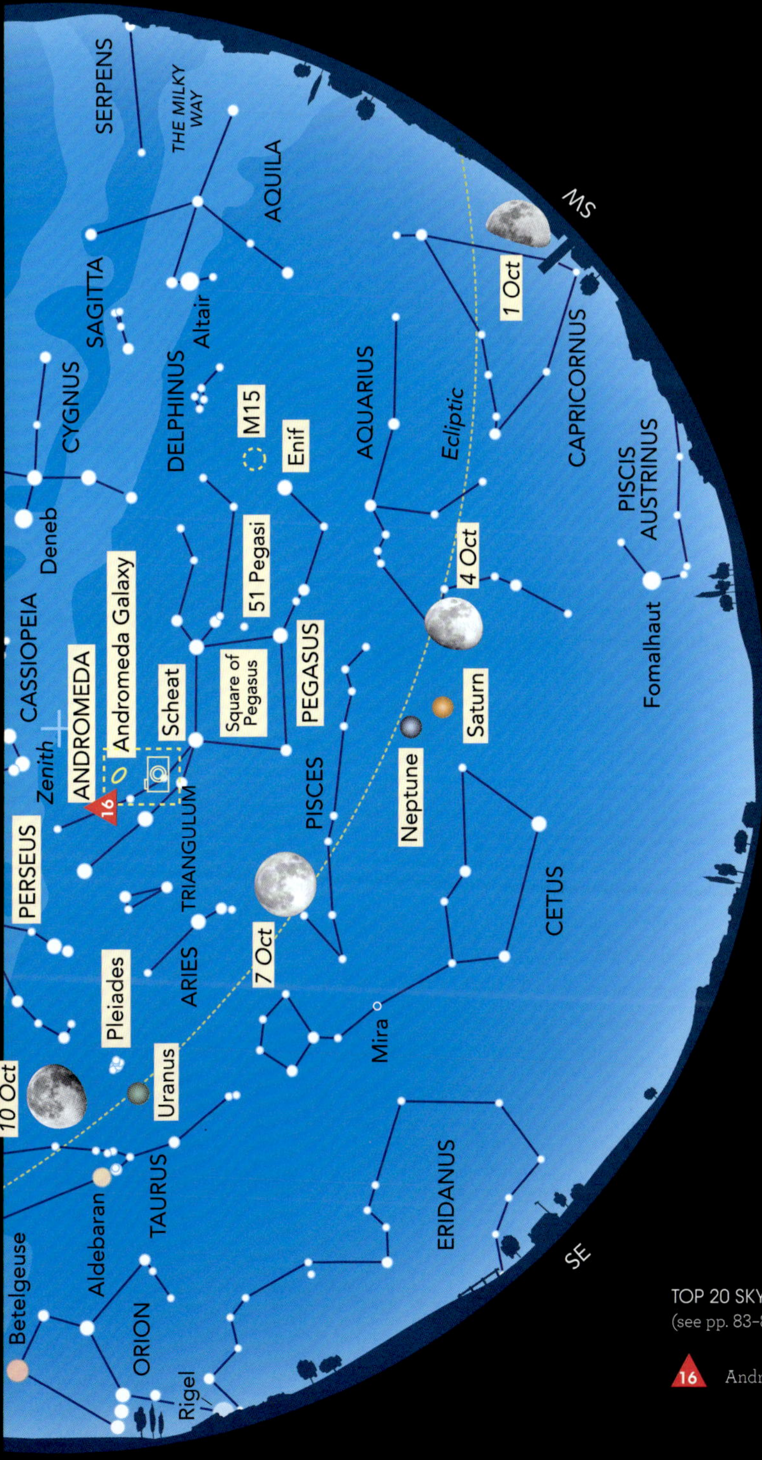

LOOKING SOUTH

WEST
SW
SE
EAST

SERPENS
THE MILKY WAY
AQUILA
SAGITTA
Altair
CYGNUS
DELPHINUS
M15
Enif
AQUARIUS
Ecliptic
1 Oct
CAPRICORNUS
PISCIS AUSTRINUS
Deneb
CASSIOPEIA
Andromeda Galaxy
51 Pegasi
4 Oct
Fomalhaut
ANDROMEDA
Scheat
Square of Pegasus
PEGASUS
Saturn
Neptune
Zenith
PISCES
PERSEUS
TRIANGULUM
CETUS
7 Oct
Pleiades
ARIES
Mira
10 Oct
Uranus
ERIDANUS
Aldebaran
TAURUS
Betelgeuse
ORION
Rigel

TOP 20 SKY SIGHTS
(see pp. 83–85)

16 Andromeda Galaxy

The constellations this month are pretty drab: neither the **Square of Pegasus** nor **Andromeda** is guaranteed to thrill. But look to the east, where the brilliant lights of winter are starting to appear, spearheaded by the beautiful star cluster of the **Pleiades**.

OCTOBER'S CONSTELLATION

Though **Pegasus** appears as little more than a large, empty square of four medium-bright stars, our ancestors managed to see it as an upside-down winged horse. In legend, Pegasus sprang from the blood of Medusa the Gorgon when Perseus severed her head.

Enif ('the nose') is an orange supergiant, with a faint blue companion visible in a small telescope. It's nicknamed the Pendulum Star: if you tap your telescope so that Enif itself seems to move up and down, you'll see the companion swinging like a pendulum (an example of an optical illusion called 'the Pulfrich effect').

A red giant star 100 times wider than the Sun, **Scheat** is nearing the end of its life and varies in brightness slightly as it pulsates. The beautiful globular cluster **M15** is a swarm of 100,000 stars lying some 36,000 light years away.

And Pegasus contains the first planet to be discovered beyond our Solar System, orbiting the star **51 Pegasi**, which is just visible to the unaided eye. The planet has been named Dimidium (meaning 'half' in Latin, as it's half the mass of Jupiter).

OCTOBER'S OBJECT

Take advantage of autumn's dark nights to pick out the Milky Way's giant neighbour, the **Andromeda Galaxy** (catalogued as M31 after its ranking in Charles Messier's list of fuzzy patches).

Lying 2.5 million light years away, Andromeda is a spiral galaxy much like our own. It's a little wider than the Milky Way, and contains twice as many stars.

From Whittington in Shropshire, Pete Williamson captured this exquisite view of the Andromeda Galaxy using a Takahashi 90-mm f/5 refractor and an FLI camera. He combined three 60-minute exposures, taken through filters passing red, green and blue light.

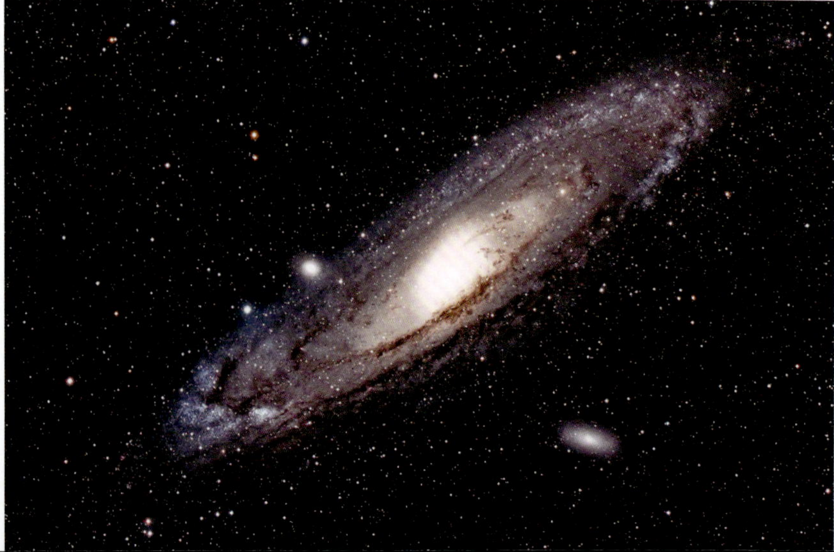

But the total masses of these two star cities are practically the same, with the scales perhaps slightly tipped in favour of the Milky Way, showing that our Galaxy is more abundantly filled with invisible 'dark matter' than our companion.

While other galaxies are receding from us in the expanding Universe, Andromeda is approaching the Milky Way, on course for a cosmic collision (see January's Topic). The outer fringes of the two galaxies – extended 'haloes' of gas surrounding the starry spirals – are already touching. The main bodies of the galaxies will merge in about 4 billion years' time to create a giant elliptical galaxy, nicknamed Milkomeda.

OBSERVING TIP

The Andromeda Galaxy is often described as the furthest object 'easily visible to the unaided eye'. It can be elusive, though – especially if you are suffering from light pollution. The trick is to memorise Andromeda's pattern of stars, and then to look slightly to the *side* of where you expect the galaxy to be. This technique – called 'averted vision' – causes the image to fall on the outer region of your retina, which is more sensitive to light than the central region that's evolved to discern fine details. Averted vision is also crucial when you want to observe the faintest nebulae or galaxies through a telescope.

OCTOBER'S TOPIC: PRECESSION

The first star most of us can identify is the North Star, otherwise known as the Pole Star or **Polaris**. Find it using the Plough (see July's Constellation) and you know you are looking due north. Polaris remains practically stationary in the sky as our planet spins, because it's almost directly above the Earth's North Pole.

The Great Pyramid at Giza, in Egypt, was built so that its sides ran precisely south to north. But the ancient Egyptians didn't use Polaris in their alignment. In their time – 4500 years ago – **Thuban**, in **Draco** (the Dragon) lay over the North Pole.

In fact, there's an endless procession of 'north stars', because the Earth's axis is not fixed in space. Instead, it sweeps around in a large circle, like the axis of a spinning top that's about to fall over – but in extreme slow motion, taking almost 26,000 years to complete one circle in the sky. This 'precession of the equinoxes' was discovered by the ancient Greek astronomer Hipparchus in 127 BC.

Currently, there's no significant star marking the south pole of the sky. The nearest candidate, sigma Octantis, is barely visible to the naked eye. But hang on 12,000 years, and brilliant Canopus – the second brightest star in the sky – will form a resplendent South Star.

Around the same time, the northern Pole Star will be the fourth-brightest star **Vega**. Navigators then – if they have to rely on star navigation – will be guided by a fitting pair of cosmic beacons.

OCTOBER'S PICTURE

The true majesty of our giant neighbour the Andromeda Galaxy – its extensive arms covering four Moon-widths of sky – is well displayed in Pete Williamson's fine photograph. In addition to showing details in the dusty spiral arms, where new stars are being born, the image reveals the two biggest of Andromeda's 30-odd companion galaxies: M32 to the left, and M110 on the lower right.

OCTOBER'S CALENDAR

SUNDAY	MONDAY	TUESDAY	WEDNESDAY	THURSDAY	FRIDAY	SATURDAY
			1	2	3	4
5 Moon near Saturn	6 Moon near Saturn (am)	7 4.47 am Full Moon; supermoon	8 Draconids (am)	9	10 Moon occults the Pleiades (am)	11
12	13 7.13 pm Last Quarter Moon near Jupiter	14 Moon near Jupiter (am)	15	16	17 Moon near Regulus (am)	18
19 Moon near Venus (am)	20	21 1.25 pm New Moon; Orionids	22 Orionids (am)	23	24	25
26 BST and IST end; Moon near Regulus (am)	27	28	29 4.21 pm First Quarter Moon; Mercury E elongation	30	31	

SPECIAL EVENTS

- **Night of 5/6 October:** the Moon passes over Saturn.
- **7 October:** the first of this year's three supermoons (see November's Special Events).
- **8 October, before dawn:** the usually sparse **Draconid meteor shower** may send some bright fireballs our way this morning, as our planet is impacted by dust particles shed by Comet Giacobini–Zinner between 1907 and 1953.
- **10 October, 6–8 am:** the Moon occults the Pleiades as the dawn sky brightens (best seen from the north of Scotland).
- **Night of 13/14 October:** you'll find Jupiter below the Last Quarter Moon, with Castor and Pollux above (Chart 10a).
- **19 October, before dawn:** the crescent Moon lies to the upper right of Venus (Chart 10b).
- **Night of 21/22 October:** it's an excellent year for observing the **Orionid meteor shower** – high-speed debris from Halley's Comet smashing into Earth's atmosphere – as the Moon is well out of the way.
- **26 October, 2 am:** the end of British Summer Time and Irish Standard Time.

OCTOBER'S PLANET WATCH

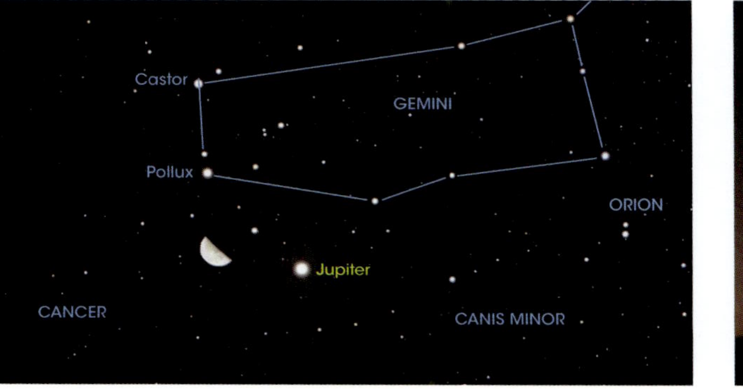

10a *14 October, 4 am. The Moon lies near Jupiter, Castor and Pollux.*

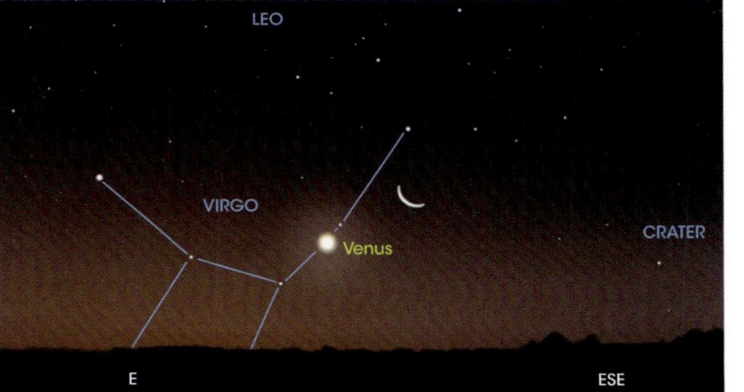

10b *19 October, 6.30 am. The crescent Moon pairs up with Venus.*

- Among the dim stars on the border of Pisces and Aquarius, **Saturn** (magnitude +0.7), sets around 5 am. The Moon glides over the ringworld on the night of 5/6 October.
- **Neptune** lies a few degrees to the left of Saturn, swimming among the undistinguished stars of Pisces. At a dim magnitude +7.7, the outermost planet sets about 5.30 am.
- Rising around 7.30 pm, **Uranus** (magnitude +5.6) is in Taurus, below the Pleiades.
- You'll find giant planet **Jupiter** in Gemini this month, below the constellation's twin leading lights, Castor and Pollux. Shining at magnitude –2.2, Jupiter rises around 11 pm. The Last Quarter Moon travels over Jupiter (and beneath the twin stars) on the night of 13/14 October (Chart 10a).
- **Venus** rises at 5.30 am, resplendent in the east at magnitude –3.9. The Moon forms a lovely pairing with the Morning Star on the morning of 19 October (Chart 10b).

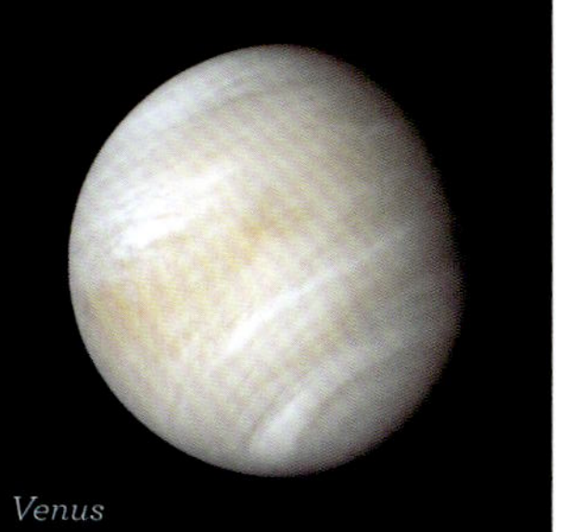

Venus

- **Mars** is lost in the Sun's glare this month, as is **Mercury** – even though the innermost planet is at its greatest separation from the Sun on 29 October.

- The sky at 10 pm in mid-November, with Moon positions at three-day intervals either side of Full Moon.
- The star positions are also correct for 11 pm at the beginning of November, and 9 pm at the end of the month.
- The planets move slightly relative to the stars during the month.

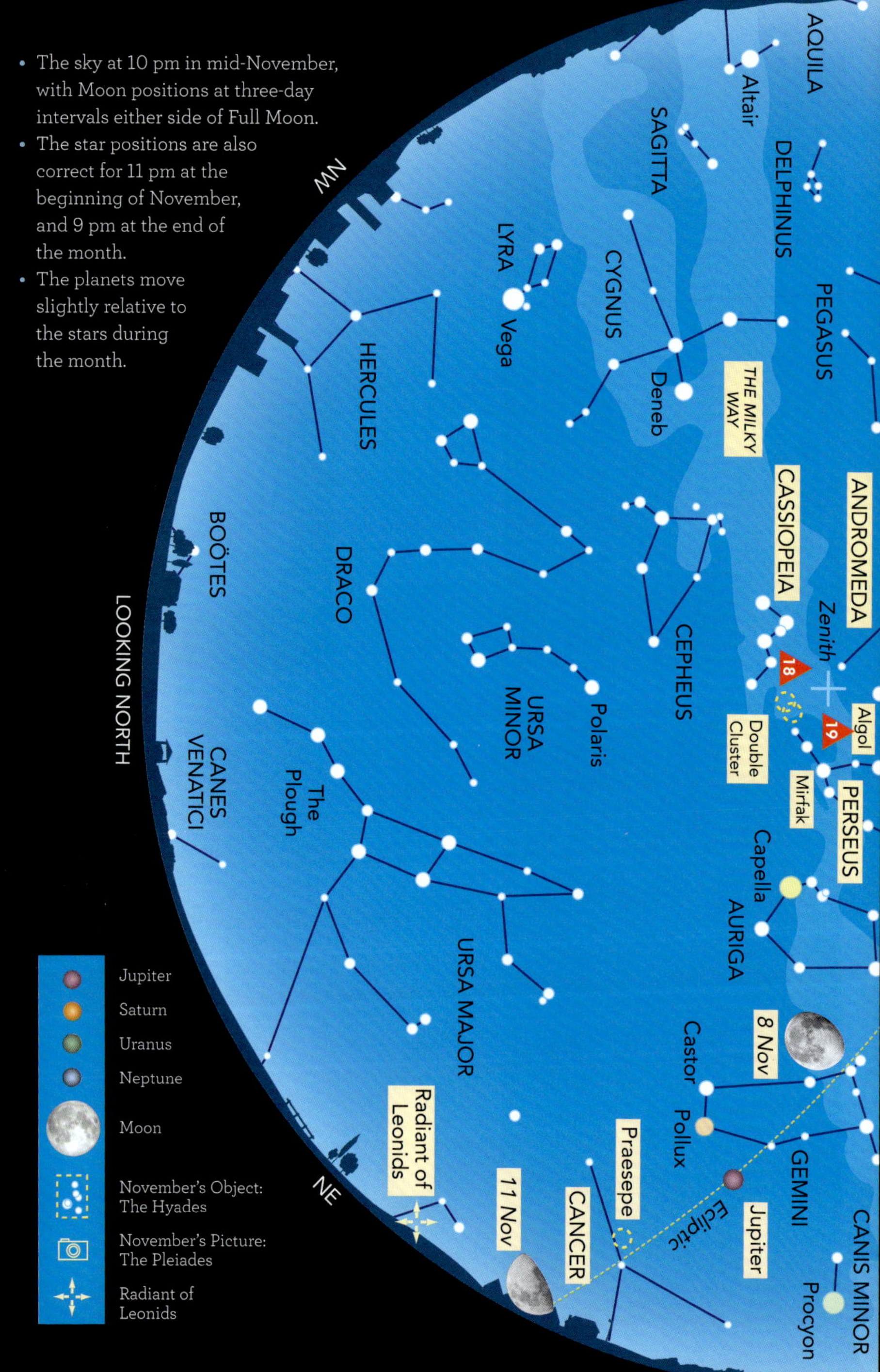

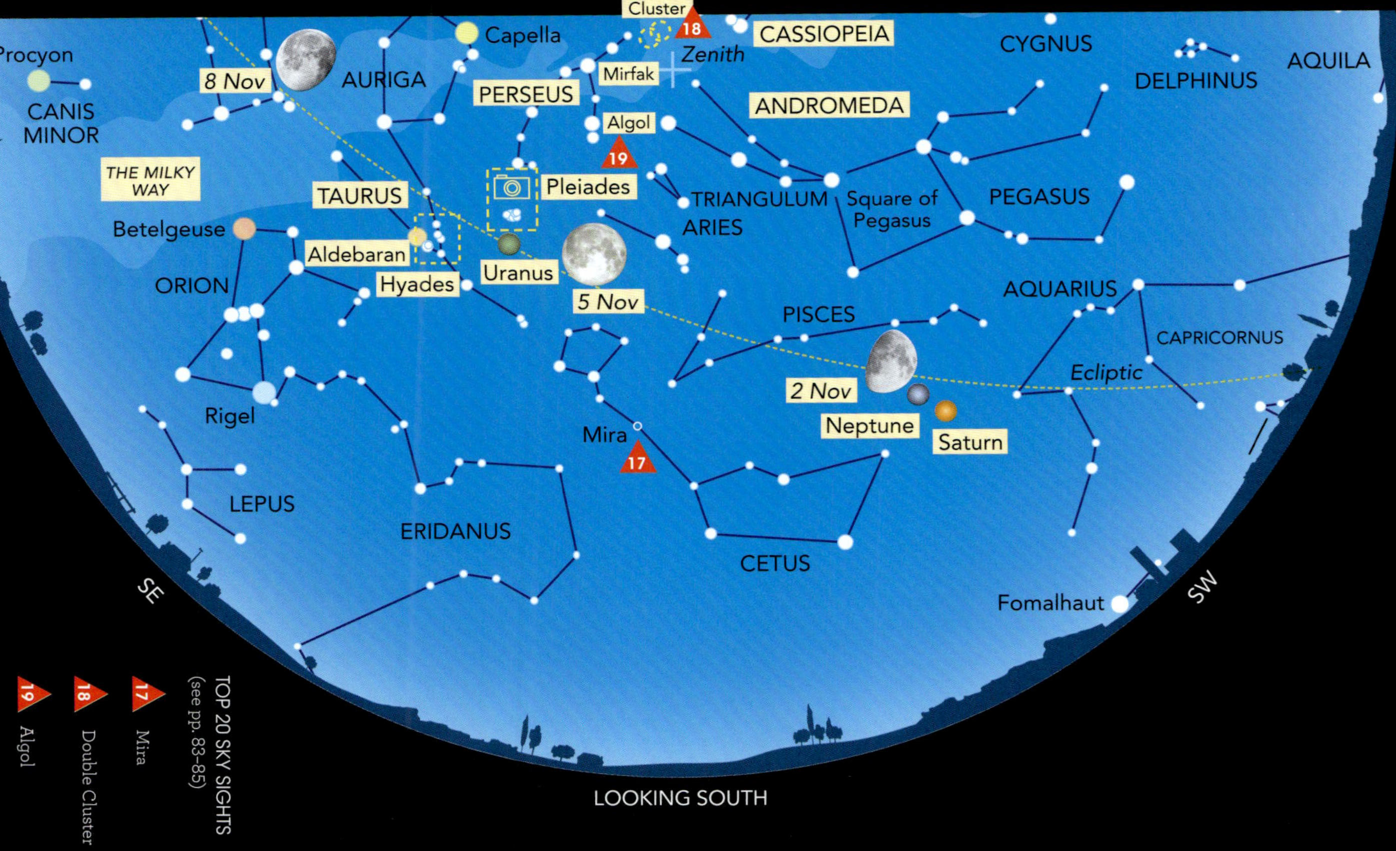

TOP 20 SKY SIGHTS
(see pp. 83–85)

- 17 Mira
- 18 Double Cluster
- 19 Algol

NOVEMBER

This month sees the biggest and brightest Full Moon of the year. There's a more subtle beauty in the **Milky Way**, rearing overhead: sweep along the glowing band with binoculars, and you'll find it studded with star clusters and nebulae.

NOVEMBER'S CONSTELLATION

One of the great heroes of Greek mythology, **Perseus**, slew the fearsome Medusa – whose gaze turned men to stone – on his way to rescue Princess **Andromeda**. The star **Algol** represents Medusa's severed head in Perseus's hand, its name stemming from the Arabic *al-Ghul* ('the demon'). Algol dims from magnitude +2.1 to +3.4 and back over a period of 2 days 21 hours. The reason was first deduced by 18-year-old John Goodricke, a profoundly deaf amateur astronomer, who correctly surmised that a dimmer star is in orbit around a more brilliant star and periodically eclipses it.

The constellation's brightest star, **Mirfak** ('elbow' in Arabic), lies 590 light years away and is 5000 times more luminous than our Sun. It's surrounded by a gaggle of slightly fainter stars making up the Alpha Persei cluster.

OBSERVING TIP

With Christmas on the way, you may be thinking of buying a telescope as a present for a budding stargazer. Beware! Unscrupulous websites and mail-order catalogues often advertise small telescopes that boast huge magnifications, but all they do is show you a large blurry image. To see planets and stars sharply, you need to use a magnification no more than twice the diameter of the lens or mirror in millimetres. So if you see an advertisement for a 75-mm telescope, beware of any claims for a magnification greater than 150 times.

On the border with **Cassiopeia** you'll find a lovely pair of star clusters, h and chi Persei, together known as the **Double Cluster.** Visible to the unaided eye on a dark night, the duo is a sensational sight in binoculars. A mere 14 million years old, the clusters are thronged with bright young blue-white stars, but they are seriously dimmed by their immense distance – 7500 light years. If the Double Cluster were as close to us as Mirfak, they would fill a region as large as Cassiopeia with hundreds of celestial jewels.

NOVEMBER'S OBJECT

According to Greek legend, the V-shaped **Hyades** star cluster, forming the 'head' of **Taurus** (the Bull), was a group of nymphs who cared for the wine god Bacchus as a baby. To the Romans, these stars were little pigs, while the Chinese saw them as a net full of rabbits.

The Hyades is the nearest star cluster to the Earth, lying 153 light years away. With the unaided eye, you can make out about 15 stars in this group while a large telescope reveals a population of several hundred fainter stars. (Although **Aldebaran** looks as though it's in the Hyades, this red giant lies at only half the distance.)

With an age of 625 million years, the Hyades is quite young on the cosmic scale. And it has a twin: **Praesepe** – the Beehive Cluster in **Cancer** – is the same age and moves in the same direction, so the two were most likely born together.

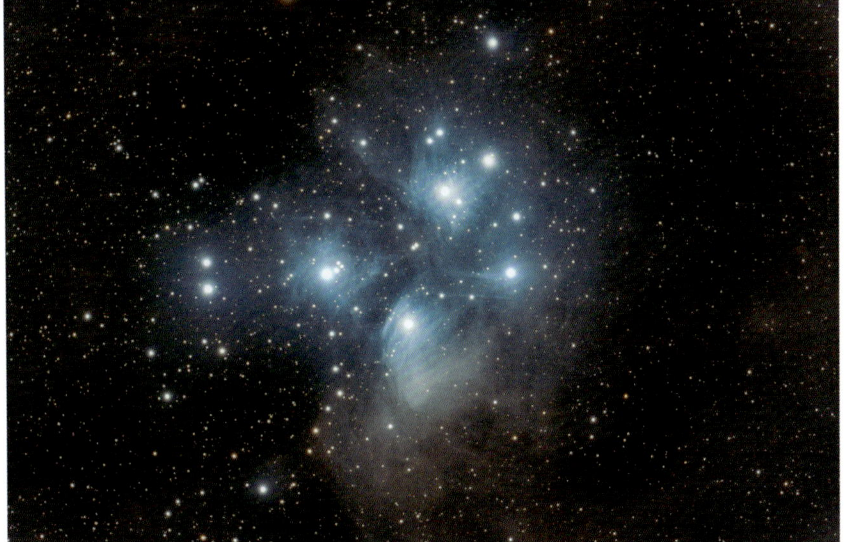

Tracey Snelus shot this lovely view of the Pleiades from North Wales. Her telescope was a William Optics GT81 81-mm f/5.9 refractor equipped with a cooled ZWO ASI071 one-shot colour camera. Over two nights and a total observing time of 8 hours 35 minutes, she took exposures of 49 × 300 seconds, 6 × 600 seconds and 70 × 180 seconds, and combined them in DeepSkyStacker.

NOVEMBER'S TOPIC: JOHN MICHELL (1724–1793)

The idea of black holes in the Universe wasn't the brainchild of Stephen Hawking, or even Albert Einstein: the idea dates back almost 250 years, to a clergyman living in Thornhill, a small village in Yorkshire...

Educated at Cambridge, John Michell was a polymath, proficient in ancient Greek and Hebrew as well as being their best scientist since Isaac Newton. But he largely allowed others to take credit for his ground-breaking ideas, including a method for weighing the Earth, a theory of earthquakes and a way of measuring the distances to the stars.

In November 1783, Michell calculated that the gravity of a very massive star – around 100 million times heavier than the Sun – would be strong enough to pull back its own light. 'If there should really exist in nature any [such] bodies . . . their light could not arrive at us'. You really can't get a much better description of a black hole!

The only nuance astronomers would add today is that a black hole doesn't have to be so heavy. Instead, you can intensify a star's gravity by compressing it down to a smaller size. That's how our Galaxy can contain black holes just a few times heavier than the Sun.

NOVEMBER'S PICTURE

The Seven Sisters visible to the naked eye comprise just a fraction of the 1000 stars making up the glittering **Pleiades** star cluster. They are a delightful sight in binoculars, and magnificent when you view them through a wide-field telescope.

As Tracey Snelus's image shows, the stars are hot and blue – fledglings on the celestial age scale – and surrounded by glorious nebulosity. Rather than material left over from the stars' birth, these glowing tendrils are part of an interstellar dust cloud which the Pleiades is currently crashing through.

NOVEMBER'S CALENDAR

SUNDAY	MONDAY	TUESDAY	WEDNESDAY	THURSDAY	FRIDAY	SATURDAY
30						1 Moon near Saturn
2 Moon near Saturn	3	4	5 1.19 pm Full Moon; supermoon	6 Moon near the Pleiades	7	8
9 Moon near Jupiter	10 Moon near Jupiter (am)	11	12 5.28 am Last Quarter Moon	13 Moon near Regulus (am)	14	15
16	17 Moon near Spica (am); Leonids	18 Leonids (am); Moon near Venus (am)	19	20 6.47 am New Moon	21 Uranus opposition	22
23	24	25	26	27	28 6.59 am First Quarter Moon	29 Moon near Saturn

SPECIAL EVENTS

- **1 November:** the 'star' to the left of the Moon is Saturn.
- **2 November:** Saturn lies to the right of the Moon.
- **5 November:** the biggest and brightest of the three supermoons this year, and the best since 2019. The Full Moon lies just 356,980 km from the Earth, and appears 14% larger and 30% more brilliant than the faintest Full Moon.
- **6 November:** as it grows dark, you'll spot the Pleiades just to the right of the Moon; with optical aid, you can also make out Uranus to the lower right (Chart 11a).
- **Night of 9/10 November:** the Moon passes above Jupiter and below Castor and Pollux (see Planet Watch and Chart 11b).
- **Night of 17/18 November:** the maximum of the **Leonid meteor shower**, fragments of Comet Tempel–Tuttle impacting our atmosphere at high velocity. This year enjoy the shooting stars under a dark moonless sky.
- **18 November, before dawn:** the crescent Moon lies to the right of Venus, low in the morning twilight.
- **21 November:** Uranus is at its closest to the Earth – 2,769 million km away – and opposite to the Sun in the sky (see Planet Watch).
- **29 November:** Saturn lies below the Moon.

NOVEMBER'S PLANET WATCH

11a 6 November, 6 pm. The Moon with the Pleiades and Uranus.

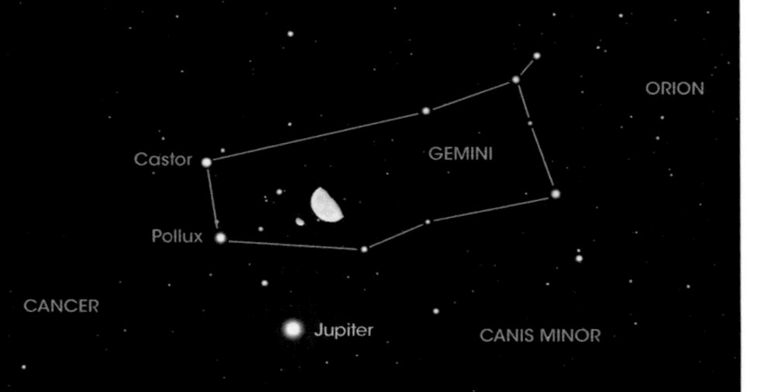

11b 9 November, 11 pm. The Moon with Jupiter, Castor and Pollux.

- You'll find **Saturn** over in the south-west during the evening, outshining all the stars in its vicinity at magnitude +0.8. The ringworld lies in Aquarius, and sets about 2 am. The Moon is nearby on 1 and 2 November; and then again on 29 November.
- Dim **Neptune** skulks in the neighbouring constellation of Pisces, at a mere magnitude +7.7 and setting around 2.30 am.
- **Uranus** is in Taurus, below the Pleiades, and reaches its closest point to Earth on 21 November (see Special Events): even so, it's barely visible to the naked eye, at magnitude +5.6. On that date, the seventh planet is also opposite the Sun in the sky and lies above the horizon all night long. The Moon is nearby on 6 November (see Special Events and Chart 11a).
- Through a small telescope, you may be able to make out the planet's tiny greenish-blue disc; but you'll need an instrument with an aperture of at least 200 mm to spot even the largest of Uranus's 28 moons.
- Rising magnificently in the north-east about 8 pm, **Jupiter** lies near the bright stars of Gemini, Castor and Pollux. The giant planet shines at magnitude –2.4. The Moon is nearby on the night of 9/10 November (Chart 11b).
- After its stint as Morning Star for the past eight months, **Venus** bows out in November as it sinks into the dawn twilight. Brilliant to the last, at magnitude –3.9, the planet is rising at around 6 am.
- On the last couple of mornings of November, Venus's place is taken by **Mercury**, rising about 6 am and shining 50 times more faintly than Venus at magnitude +0.3.
- **Mars** appears too close to the Sun to be seen in November.

- The sky at 10 pm in mid-December, with Moon positions at three-day intervals either side of Full Moon.
- The star positions are also correct for 11 pm at the beginning of December, and 9 pm at the end of the month.
- The planets move slightly relative to the stars during the month.

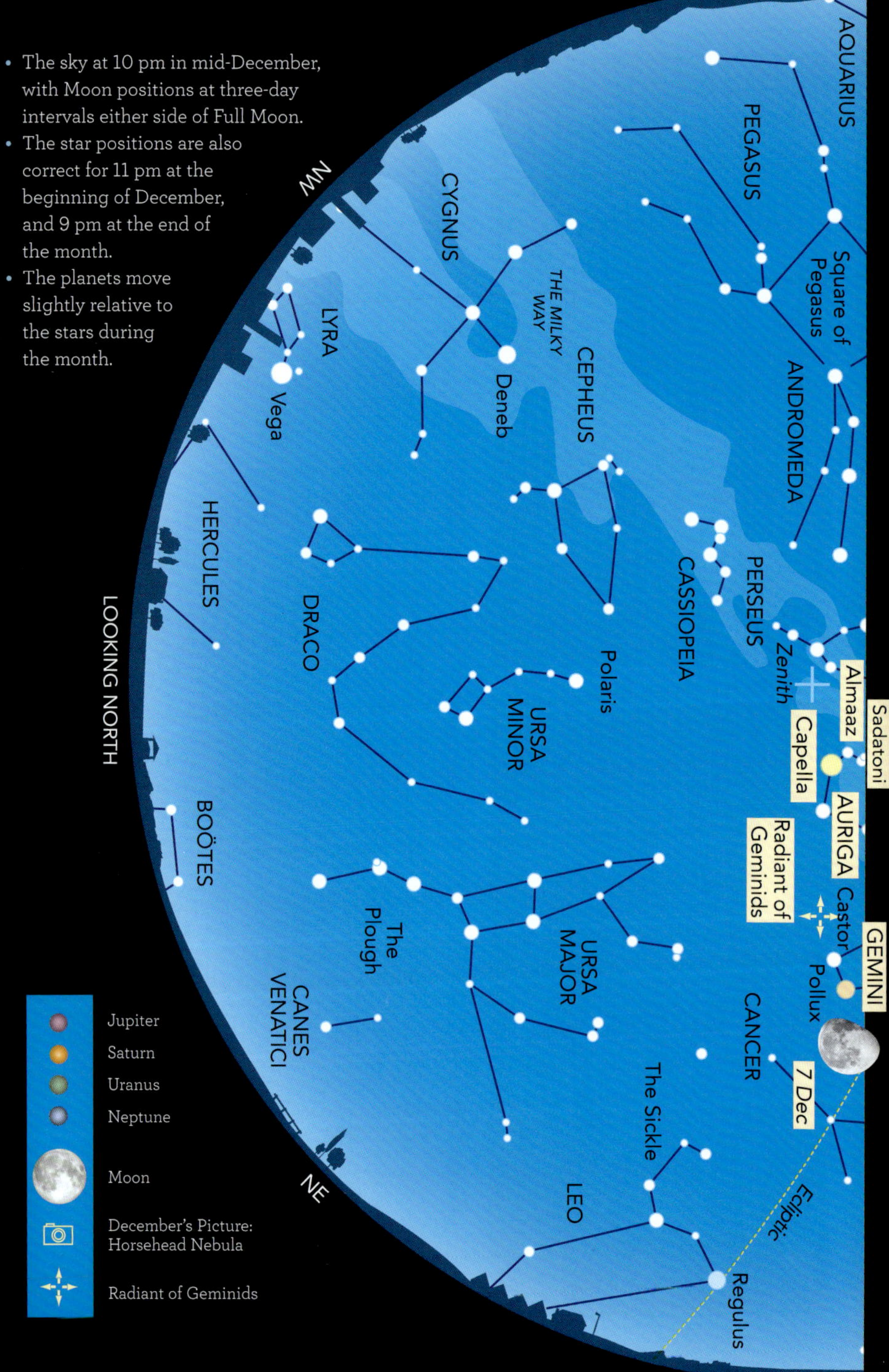

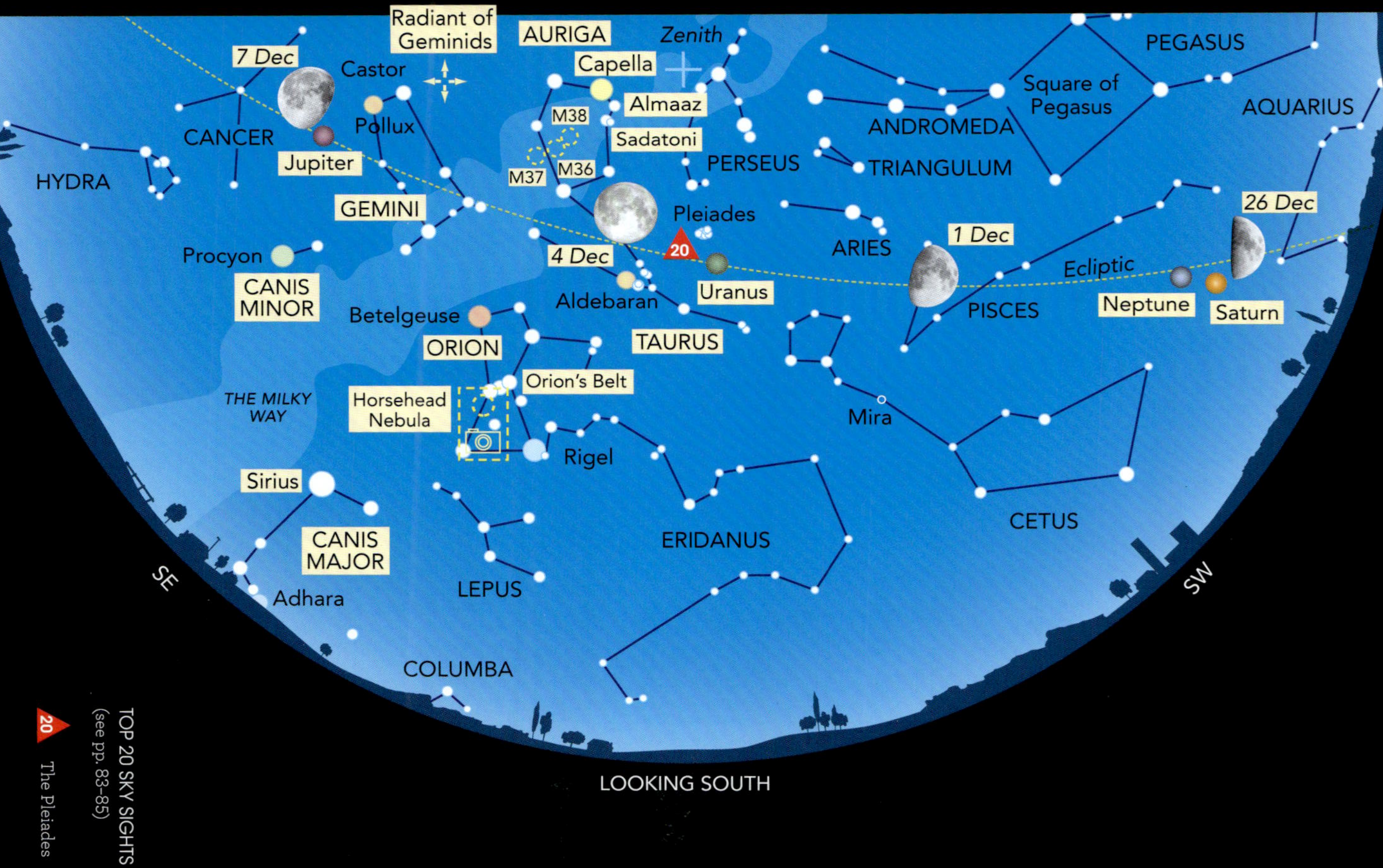

TOP 20 SKY SIGHTS
(see pp. 83–85)

20 The Pleiades

DECEMBER

Look out for a brilliant meteor shower, plus the Moon playing hide-and-seek with Regulus and with the Pleiades. The stars, too, are putting on a celebratory show, featuring **Orion** with his hunting dogs **Canis Major** and **Canis Minor**, co-starring **Taurus** (the Bull), the hero twins of **Gemini** and the ancient speedway driver **Auriga**.

DECEMBER'S CONSTELLATION

Sparkling overhead, **Auriga** (the Charioteer) is named after the Greek hero Erichthonius, who invented and raced the four-horse chariot. The ancient Babylonians, on the other hand, saw Auriga as a shepherd's crook.

Capella, the sixth-brightest star in the sky, means 'the little nanny goat'. It actually consists of two stars, each 75 times brighter than the Sun, orbiting more closely than the Earth circles the Sun.

Nearby, you'll find two eclipsing binaries: stars that regularly change in brightness because a companion passes in front. **Sadatoni** is an orange star eclipsed every 2 years 8 months by a blue partner. **Almaaz** has a companion star that's completely hidden by a huge dark disc of cosmic dust swirling around it. Every 27 years, this disc moves in front of Almaaz and eclipses it for two years.

OBSERVING TIP

Hold a meteor party to check out the Geminid meteor shower on the night of 13/14 December. You don't need any optical equipment: the ideal viewing equipment is your unaided eye, plus a warm sleeping bag and a lounger. Everyone should look in different directions, so you can cover the whole sky. Shout out 'meteor!' when you see a shooting star. One of your party can record the observations, using a watch, notepad and red torch.

And bring out your binoculars (better still, a small telescope) to view three very pretty star clusters within the outline of Auriga, **M36**, **M37** and **M38**.

DECEMBER'S OBJECT

Tiny Mercury is the 'Planet of the Month'. Not only can you see it in the morning sky, soon after 6 am (see Planet Watch), but a spacecraft named BepiColombo is scheduled to slip into orbit and scrutinise the planet in unprecedented detail.

The innermost world in the Solar System, Mercury never strays far from the Sun in the sky. Whizzing round its tiny oval-shaped orbit, Mercury completes one 'year' in just 88 Earth-days. In contrast, the planet rotates so extremely slowly on its axis that its 'day' from sunrise to sunrise is twice as long as Mercury's year!

Previous robotic spacecraft have discovered that Mercury – only a little bigger than our Moon – is also peppered by craters. But that's where the resemblance ends. Mercury's surface has wrinkled like the skin of a dried-up apple as the planet's supersized iron core has shrunk.

Astronomers hope BepiColombo will tell us why Mercury has such a large metallic core: was the planet once much larger, but had its rocky layers blasted off when it was hit by another wayward world? And why does Mercury have a magnetic field more powerful than the larger planets Venus and Mars?

Among the mysteries of Mercury's surface under BepiColombo's scrutiny will be patches of frozen ice near the poles of a planet where the temperature near the equator rises to 430°C.

DECEMBER'S TOPIC: WHITE DWARFS

Brilliant **Sirius** (see February's Object) is accompanied by a most extraordinary object. First predicted by its gravitational pull on the Dog Star, this small companion star is often fondly nicknamed 'the Pup'. At magnitude +8.4, the Pup should be visible in binoculars. But Sirius glares 10,000 times brighter at magnitude –1.5, and you'll need a 150-mm telescope to spot its small companion.

The Pup is only the size of the Earth, yet it boasts 98 per cent the mass of our Sun. Hotter than the Sun, it shines a pure white colour – hence the moniker 'white dwarfs' for these objects, which make up 10 per cent of all the stars in the Galaxy.

A white dwarf is the core of an old star that has died, having puffed off its atmosphere as a brief – but glorious – planetary nebula. The core has collapsed under its own gravity, squeezing electrons and the nuclei of atoms together to unbelievable densities: a matchbox full of white-dwarf material would outweigh an elephant!

With no power source, a white dwarf eventually just fades away, to end up as a cold, black cinder. Seven billion years hence, this is the fate that awaits our Sun.

DECEMBER'S PICTURE

The celestial chess piece of our Galaxy, the **Horsehead Nebula** lies just below **Orion's Belt**. Seven light years from nose to mane, it's a dark cloud of dust and gas, silhouetted against bright nebulosity

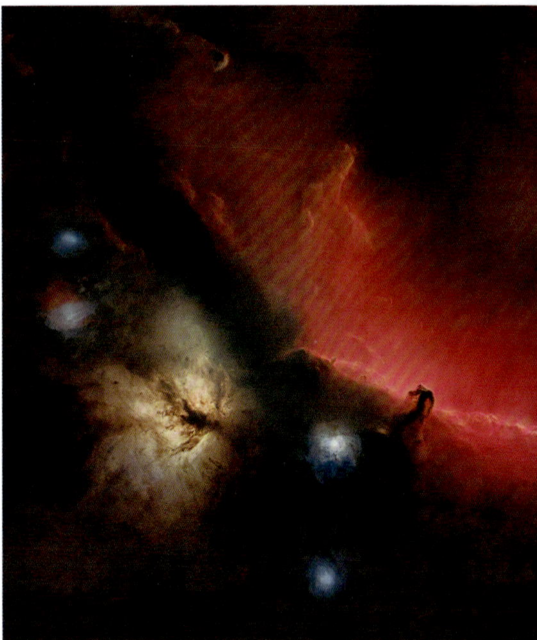

Using an Askar FRA600 108-mm refractor and QHY268M camera, Simon Hudson took a series of long exposures through Astronomik narrowband and LRGB filters to capture the Horsehead Nebula and its environs. They comprised 50 × 300 seconds in hydrogen alpha, 36 × 300 seconds luminance, 28 × 300 seconds in red, 21 × 300 seconds in green and 30 × 300 seconds in blue: the total exposure time was almost 14 hours. He used PixInsight and Photoshop software to process the image and remove the stars.

that's lit up by hot young stars.

The Horsehead is notoriously difficult to spot through a telescope, because the background nebulosity is so dim: you'll probably need a 300-mm instrument, zero light pollution and 20 minutes adapting your eyes to the dark.

But it's a magnificent sight in photos, as exemplified here by Simon Hudson's masterpiece. He has deliberately removed all the stars from this view, to reveal the intricate details of the Horsehead, the Flame Nebula (to the left) and the backdrop of swirling fronds of gas.

DECEMBER'S CALENDAR

SUNDAY	MONDAY	TUESDAY	WEDNESDAY	THURSDAY	FRIDAY	SATURDAY
	1	2	3 Moon near Pleiades and Uranus	4 Moon occults Pleiades (am); 11.14 pm Full Moon; supermoon	5	6
7 Mercury W elongation; Moon near Jupiter	8	9 Moon near Regulus	10 Moon occults Regulus (am)	11 8.52 pm Last Quarter Moon	12	13 Geminids
14 Geminids (am); Moon near Spica (am)	15	16	17 Moon near Mercury (am)	18	19	20 1.43 am New Moon
21 Winter Solstice	22	23	24	25	26 Moon near Saturn	27 7.10 pm First Quarter Moon
28	29	30	31 Moon near Pleiades			

SPECIAL EVENTS

- **Night of 3/4 December:** the Moon moves right in front of the Pleiades between 3 and 5.30 am, occulting most of its brightest stars.
- **4 December:** the last of this year's three supermoons (see November's Special Events).
- **7 December, before dawn:** Mercury reaches its greatest separation from the Sun (see Planet Watch).
- **7 December:** Jupiter is near the Moon, Castor and Pollux.
- **9 December:** Regulus lies to the left of the Moon.
- **10 December, 7.20–8.20 am:** the Moon occults Regulus as seen from all of the UK and Ireland, with the exact time depending on your location. The star reappears during the dawn twilight for western Ireland and northern Scotland, but in full daylight for south-eastern England.
- **Night of 13/14 December:** dust grains from the asteroid Phaethon blaze in our atmosphere as the **Geminid meteor shower**, the most spectacular shooting star display of the year. Best seen before the Moon rises at 1 am.
- **17 December, before dawn:** Mercury lies to the left of the crescent Moon, low in the morning twilight (Chart 12b).
- **21 December, 3.03 pm:** the Winter Solstice, when the Sun reaches its southernmost point in the sky, giving the northern hemisphere the shortest day and the longest night.
- **26 December:** the waxing crescent Moon lies to the right of Saturn.

DECEMBER'S PLANET WATCH

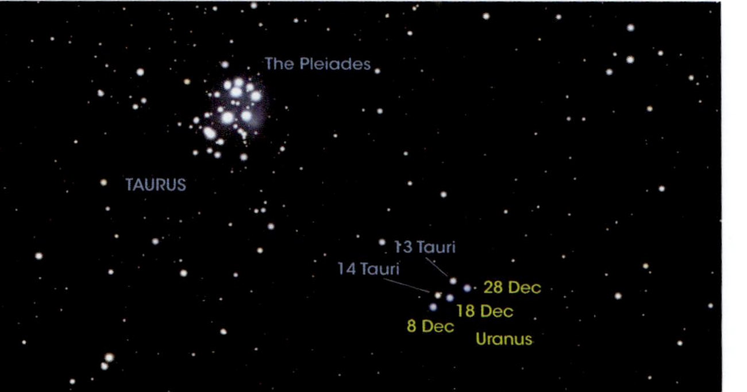

12a 8–28 December, 7 pm. Uranus passes 14 and 13 Tauri, near the Pleiades.

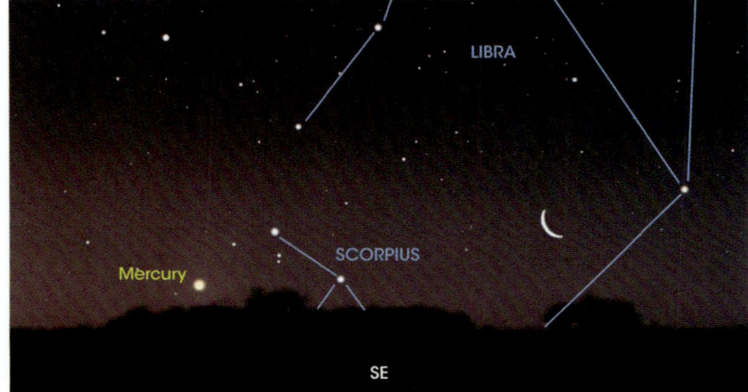

12b 17 December, 6.45 am. The crescent Moon with Mercury.

• **Saturn** is over in the south-west after darkness falls, shining at magnitude +1.0 in Aquarius and setting around midnight. The Moon is nearby on 26 December.

• In the neighbouring constellation Pisces, **Neptune** glows at a dim magnitude +7.8 and sets about 0.30 am.

• **Uranus** (magnitude +5.6) is in Taurus, some 4 degrees below the Pleiades. The seventh planet is setting about 6 am. Around the middle of the month, Uranus passes two stars – 14 and 13 Tauri – that nearly match it in brightness, forming an ever-changing triangle as seen in binoculars (Chart 12a).

• The brightest planet visible this month, **Jupiter**, rises at 6 pm at a brilliant magnitude –2.6. The giant world hangs below Castor and Pollux in Gemini. The Moon passes nearby on 7 December.

• **Mercury** puts on its best morning show of the year during the first half of December. At its greatest separation from the Sun on 7 December, the innermost planet shines at magnitude –0.4 and rises at 6 am. As Mercury drops back down towards the horizon, the Moon appears to its right on the morning of 17 December (Chart 12b).

• **Venus** and **Mars** are lost in the Sun's glare this month.

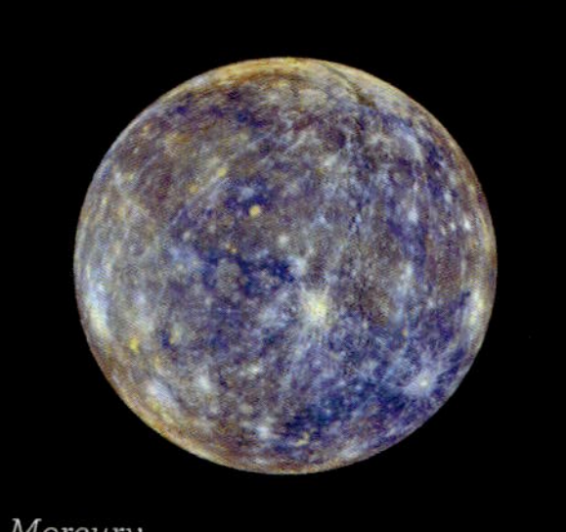

Mercury

Can you see the planets? It's a common question; and the answer is a resounding 'yes!' Some of our cosmic neighbours are the brightest objects in the night sky after the Moon. As they're so close, you can watch them getting up to their antics from night to night. And planetary debris – leftovers from the birth of the Solar System – can light up our skies as glowing comets and the celestial fireworks of a meteor shower.

THE SUN-HUGGERS

Mercury and Venus orbit the Sun more closely than our own planet, so they never seem to stray far from our local star: you can spot them in the west after sunset, or the east before dawn, but never all night long. At *elongation*, the planet is at its greatest separation from the Sun, though – as you can see in the diagram (right) – that's not when the planet is at its brightest. Through a telescope, Mercury and Venus (technically known as the *inferior planets*) show phases like the Moon – from a thin crescent to a full globe – as they orbit the Sun.

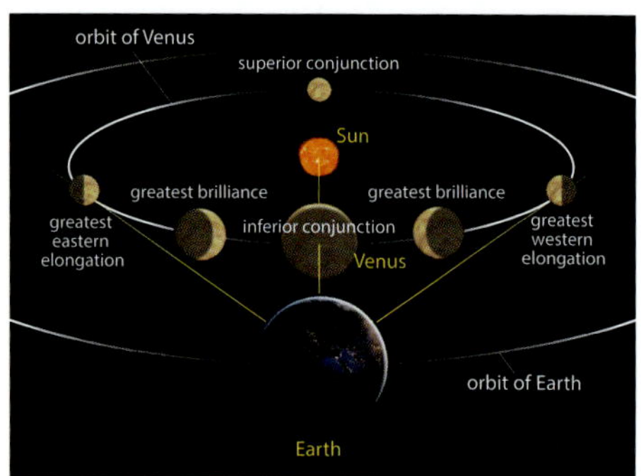

Venus (and Mercury) show phases like the Moon as they orbit the Sun.

Mercury

The innermost planet puts on its best evening performance in February–March; it's visible low down in late June, but is lost in the evening twilight at its October elongation. Before dawn, Mercury appears at the beginning of January, but is drowned out by bright twilight at its April elongation. You can catch it again at the end of August; while November–December sees Mercury at its best in the morning sky.

Maximum elongations of Mercury in 2025	
Date	**Separation**
8 March	18° east
21 April	27° west
4 July	26° east
19 August	19° west
29 October	24° east
7 December	21° west

Venus

From January to mid-March, Venus is a glorious Evening Star. On **23 March**, it swings between Sun and Earth before taking up residence as the Morning Star until November.

Maximum elongations of Venus in 2025	
Date	**Separation**
10 January	47° east
1 June	46° west

WORLDS BEYOND

A planet orbiting the Sun beyond the Earth (known in the jargon as a *superior planet*) can be visible at all times of night, as we look outwards into the Solar System. It lies due south at midnight when the Sun, the Earth and the planet are all in line – a time known as *opposition* (see the diagram, right). Around this time the Earth lies nearest to the planet, although the date of closest approach (and the planet's maximum brightness) may differ by a few days because the planets' orbits are not circular.

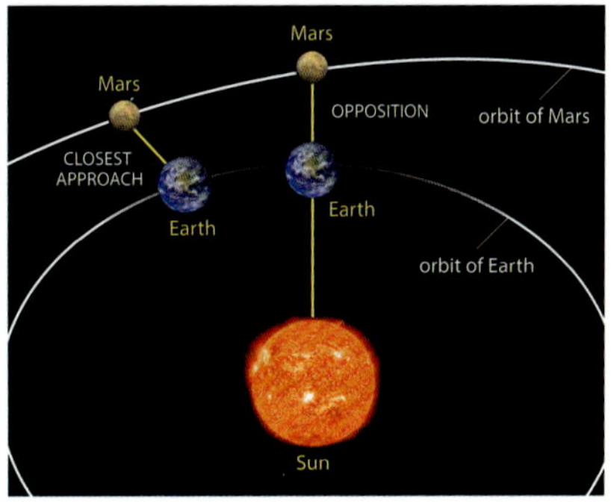

Mars (and the outer planets) line up with the Sun and Earth at opposition, but they are brightest at their point of closest approach.

Mars

The Red Planet is at its most prominent for two years at the start of 2025, closest to us on **12 January** and at opposition on **16 January**. Mars then fades rapidly as the Earth pulls away, though it remains visible in the evening sky right through to August. The planet is lost in the Sun's glow for the rest of the year.

Where to find Mars	
Early January	Cancer
Mid-January–mid-April	Gemini
Mid-April–May	Cancer
June–July	Leo
August	Virgo

Jupiter

The giant planet is brilliant in the evening sky until the end of May, lying in Taurus. Jupiter reappears before dawn in mid-July, moving into Gemini where it remains for the rest of the year. It does not reach opposition in 2025.

Saturn

In January and February you can catch Saturn to the west after sunset, in Aquarius. The Moon occults the ringworld on **4 January**. After disappearing behind the Sun, it reappears in the morning sky in May, having moved to Pisces. On **21 September**, Saturn is at opposition and at its closest to the Earth. The planet returns to Aquarius for November and December.

Uranus

The seventh planet starts 2025 in Aries. Moving to Taurus in March, Uranus hangs below the Pleiades for the remainder of the year. Up until April, Uranus is visible in the evening sky: the planet re-emerges in the morning sky in July, and reaches opposition on **21 November**.

Neptune

The most distant planet – in Pisces all year – is closest to the Earth on **22 September** and reaches opposition on **23 September**. Neptune can be seen (though only through binoculars or a telescope) in January and February and then from June until the end of the year.

SOLAR ECLIPSES

On **29 March**, a partial eclipse of the Sun is visible from north-west Europe, north-west Africa and north-east Canada,

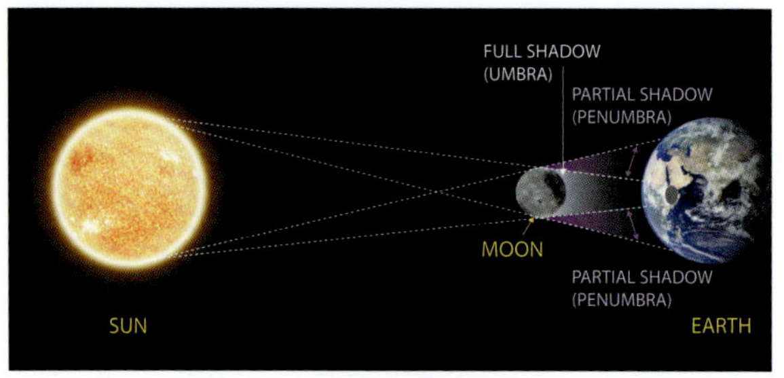

Where the dark central part (the umbra) of the Moon's shadow reaches the Earth, we are treated to a total solar eclipse. If the shadow doesn't quite reach the ground, we see an annular eclipse. People located within the penumbra observe a partial eclipse.

where 93 per cent is covered. For northern Scotland and Ireland, 45 per cent of the Sun is obscured, decreasing to 30 per cent for the south-east of England.

A partial solar eclipse can be seen on **21 September** from the southern Pacific Ocean and Antarctica, where 85 per cent of the solar disc is obscured. Nothing is visible from the UK or Ireland.

LUNAR ECLIPSES

On the morning of **14 March**, a total eclipse of the Moon is visible from the Americas and western regions of Africa and Europe. People in Ireland and the UK will witness the Moon entering totality just as it sets.

A total lunar eclipse on **7 September** is visible from Australia, Asia, Africa and Europe. As seen from the UK and Ireland, the Moon rises fully eclipsed and rapidly moves out of the Earth's shadow.

METEOR SHOWERS

Shooting stars, or *meteors,* are tiny specks of interplanetary dust burning up in the Earth's atmosphere. At certain times of year, Earth passes through a stream of debris (usually left by a comet) and we see a *meteor shower.* The meteors appear to emanate from a point in the sky known as the *radiant.* Most showers are known by the constellation in which the radiant lies.

Table of annual meteor showers

Meteor shower	Date of maximum
Quadrantids	3/4 January
Lyrids	22/23 April
Eta Aquarids	6 May (am)
Perseids	12/13 August
Orionids	21/22 October
Leonids	17/18 November
Geminids	13/14 December

It's fun and rewarding to hold a meteor party. Note the location, cloud cover, the time and brightness of each meteor and its direction through the stars – along with any persistent afterglow (train).

COMETS

Comets are dirty snowballs from the outer Solar System. If they fall towards the Sun, its heat evaporates their ices to produce a gaseous head (*coma*) and sometimes dramatic tails. Although some comets are visible to the naked eye, use binoculars to reveal stunning details in the coma and the tail.

Hundreds of comets move round the Sun in small orbits. But many more don't return for thousands or even millions of years. Most comets are now discovered in professional surveys of the sky, but a few are still found by dedicated amateur astronomers. No bright comets are expected in 2025 at the time of writing, but watch out in case a brilliant new comet puts in a surprise appearance!

Here are some of the most popular sights in the night sky, in a season-by-season summary. It doesn't matter if you're a complete beginner, finding your way around the heavens with the unaided eye or binoculars; or if you're a seasoned stargazer, with a moderate telescope. There's something here for everyone.

Each sky sight comes with a brief description, and a guide as to how you can best see it. Many of the most delectable objects are faint, so avoid moonlight when you go out spotting. Most of all, enjoy!

SPRING

Praesepe

Constellation: Cancer
Star Chart/Key: March; 5
Type/Distance: Star cluster; 600 light years
Magnitude: +3.7
A fuzzy patch to the unaided eye; a telescope reveals many of its 1000 stars.

M81 and M82 

Constellation: Ursa Major
Star Chart/Key: March; 6
Type/Distance: Galaxies; 12 million light years
Magnitude: +6.9 (M81); +8.4 (M82)
A pair of interacting galaxies: the spiral M81 appears as an oval blur, and the starburst M82 as a streak of light.

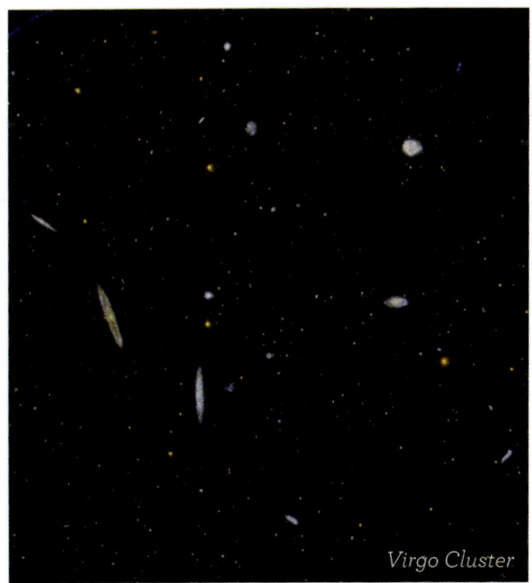
Virgo Cluster

The Plough 

Constellation: Ursa Major
Star Chart/Key: April; 7
Type/Distance: Asterism; 82–123 light years
Magnitude: Stars are roughly magnitude +2
The seven brightest stars of the Great Bear form a large saucepan shape, called 'the Plough'.

Mizar and Alcor 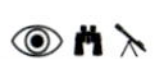

Constellation: Ursa Major
Star Chart/Key: April; 8
Type/Distance: Double star; 83 & 82 light years
Magnitude: +2.3 (Mizar); +4.0 (Alcor)
The sky's classic double star, easily separated by the unaided eye: a telescope reveals Mizar itself is a close double.

Virgo Cluster 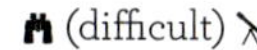

Constellation: Virgo
Star Chart/Key: May; 9
Type/Distance: Galaxy cluster; 54 million light years
Magnitude: Galaxies range from magnitude +9.4 downwards
Huge cluster of 2000 galaxies, best seen through moderate to large telescopes.

SUMMER

Antares 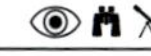

Constellation: Scorpius
Star Chart/Key: June; 10
Type/Distance: Red giant; 550 light years
Magnitude: +0.96
Bright red star close to the horizon. You can spot a faint green companion with a telescope.

M13 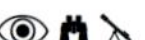

Constellation: Hercules
Star Chart/Key: June; 11
Type/Distance: Star cluster; 23,000 light years
Magnitude: +5.8
A faint blur to the naked eye, this ancient globular cluster is a delight seen through binoculars or a telescope. It boasts nearly a million stars.

Lagoon and Trifid Nebulae 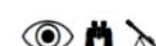

Constellation: Sagittarius
Star Chart/Key: July; 12
Type/Distance: Nebulae; 5000 light years
Magnitude: +6.0 (Lagoon); +7.0 (Trifid)
While the Lagoon Nebula is just visible to the unaided eye, you'll need binoculars or a telescope to spot the Trifid. The two are in the same binocular field of view, and present a stunning photo opportunity.

Albireo 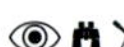

Constellation: Cygnus
Star Chart/Key: August; 13
Type/Distance: Double star; 430 light years
Magnitude: +3.2 (Albireo A); +5.1 (Albireo B)
Good binoculars reveal Albireo as being double. But you'll need a small telescope to appreciate its full glory. The brighter star appears golden; its companion shines piercing sapphire. It is the most beautiful double star in the sky.

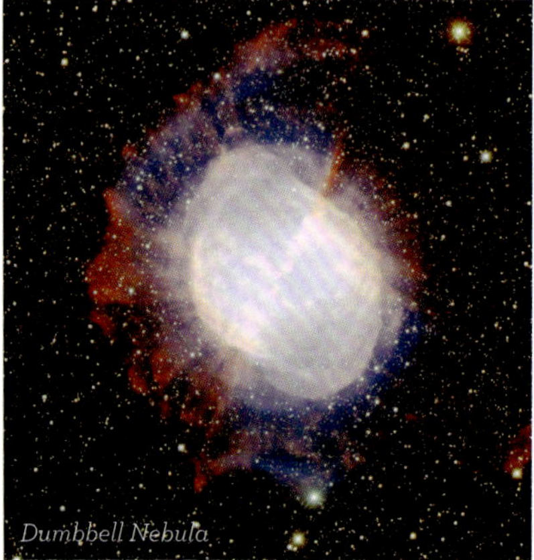
Dumbbell Nebula

Dumbbell Nebula 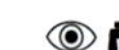

Constellation: Vulpecula
Star Chart/Key: August; 14
Type/Distance: Planetary nebula; 1270 light years
Magnitude: +7.5
Visible through binoculars, and a lovely sight through a small/medium telescope, this dying star has puffed off its atmosphere into space.

AUTUMN

Delta Cephei 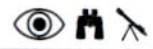

Constellation: Cepheus
Star Chart/Key: September; 15
Type/Distance: Variable star; 890 light years
Magnitude: +3.5 to +4.4, varying over 5 days 9 hours
The classic variable star, Delta Cephei is chief of the Cepheids – stars that allow us to measure distances in the Universe (their variability time is coupled to their intrinsic luminosity). Visible to the unaided eye, but you'll need binoculars for serious observations.

Andromeda Galaxy 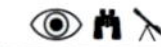

Constellation: Andromeda
Star Chart/Key: October; 16
Type/Distance: Galaxy; 2.5 million light years
Magnitude: +3.4
The nearest major galaxy to our own, the Andromeda Galaxy is easily visible to the unaided eye in unpolluted skies. Four times the width of the Full Moon, it's a great telescopic object and photographic target.

Mira

Constellation: Cetus
Star Chart/Key: November; 17
Type/Distance: Variable star; 300 light years
Magnitude: +2 to +10 over 332 days, although maxima and minima may vary.
Nicknamed 'the Wonderful', this distended red giant star is alarmingly variable as it swells and shrinks. At its brightest, Mira is a naked-eye object; binoculars may catch it at minimum; but you need a telescope to monitor this star. Its behaviour is unpredictable, and it's important to keep logging it.

Double Cluster 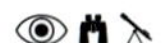

Constellation: Perseus
Star Chart/Key: November; 18
Type/Distance: Star clusters; 7500 light years
Magnitude: +3.7 and +3.8
A lovely sight to the unaided eye, these stunning young star clusters are sensational through binoculars or a small telescope. They're a great photographic target.

Algol 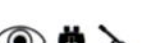

Constellation: Perseus
Star Chart/Key: November; 19
Type/Distance: Variable star; 90 light years
Magnitude: +2.1 to +3.4 over 2 days 21 hours
Like Mira, Algol is a variable star, but not an intrinsic one. It's an 'eclipsing binary' – its brightness falls when a fainter companion star periodically passes in front of the main star. It's easily monitored by the eye, binoculars or a telescope.

WINTER

Pleiades 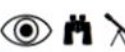

Constellation: Taurus
Star Chart/Key: December; 20
Type/Distance: Star cluster; 440 light years
Magnitude: Stars range from magnitude +2.9 downwards
To the naked eye, most people can see six stars in the cluster, but it can rise to 14 for the keen-sighted. In binoculars or a telescope, they are a must-see. Astronomers have observed 1000 stars in the Pleiades.

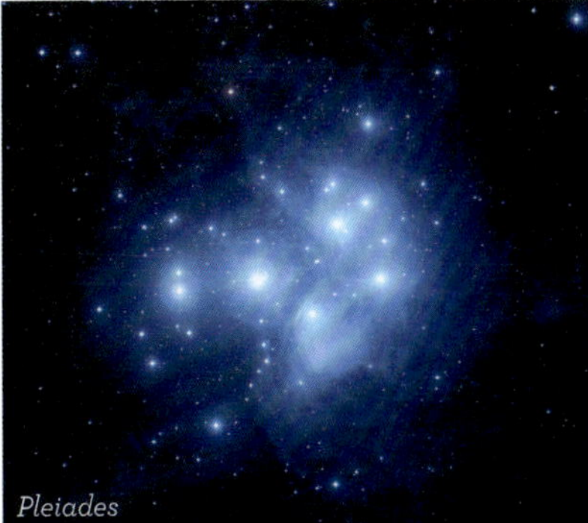

Pleiades

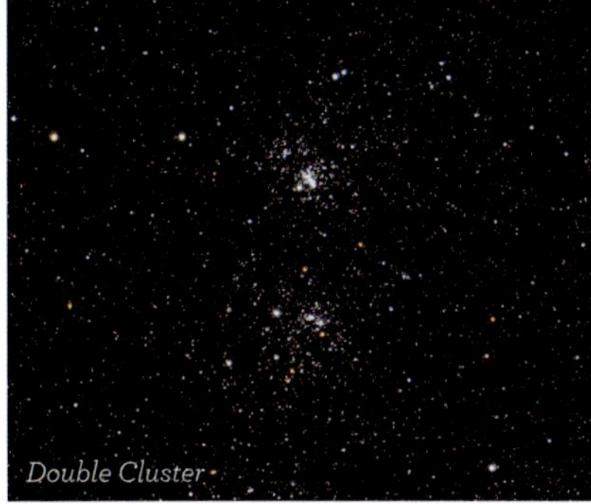

Double Cluster

Orion Nebula 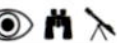

Constellation: Orion
Star Chart/Key: January; 1
Type/Distance: Nebula; 1340 light years
Magnitude: +4.0
A striking sight even to the unaided eye, the Orion Nebula – a star-forming region 24 light years across – hangs just below Orion's Belt. Through binoculars or a small telescope, it is staggering. A photographic must!

Betelgeuse 

Constellation: Orion
Star Chart/Key: January; 2
Type/Distance: Variable star; 720 light years
Magnitude: 0.0 to +1.6
Even with the unaided eye, you can see that Betelgeuse is slightly variable over months, as the red giant star billows in and out.

M35 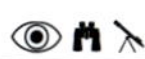

Constellation: Gemini
Star Chart/Key: February; 3
Type/Distance: Star cluster; 2800 light years
Magnitude: +5.3
Just visible to the unaided eye, this cluster of around 2000 stars is a lovely sight through a small telescope.

Sirius 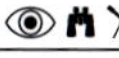

Constellation: Canis Major
Star Chart/Key: February; 4
Type/Distance: Double star; 8.6 light years
Magnitude: –1.5
You can't miss the Dog Star. It's the brightest star in the sky! But you'll need a 150-mm reflecting telescope (preferably bigger) to pick out its +8.4 magnitude companion – a white dwarf nicknamed 'the Pup'.

THE RISE OF THE SMALL REFRACTOR

BY ROBIN SCAGELL

There are more small refracting telescopes in the world than any other type. These are the traditional type of telescope, with a main lens at one end and an eyepiece at the other. For years, a small refractor was regarded as the beginner's telescope, being easy to use and quite cheap to make. The classic 'department store' telescope has been churned out in its millions, with indifferent or poor lenses, a wobbly mounting and little to recommend it as a useful instrument.

Yet small refracting telescopes are now commanding high prices among amateur astronomers. They are about the same aperture - around 40–70 mm - as the old cheap and nasty models that are lying unused in their boxes under beds across the world. But their optical performance and manufacturing quality are in a completely different league. And instead of the wobbly mountings that made finding even the Moon a challenge, they come in tube-only form, with 'dovetail' fittings that allow you to attach them to a variety of good-quality mountings. This form of supply is known as OTA, for 'optical tube assembly', and this abbreviation is widely used in the adverts for such instruments.

The Askar FMA180 is a 180-mm f/4.5 apochromatic lens mounted on a dovetail and with a saddle for a smaller finder or autoguider scope. While intended for imaging, it can be adapted so as to be used visually with an eyepiece. A wide range of APO lenses are available from various manufacturers.

Although they can be used for conventional viewing through an eyepiece, they are generally not sold with an eyepiece as they are intended mainly for imaging. So in effect these are really a form of telephoto lens, and they can give superb pictures of deep-sky objects such as nebulae and the larger galaxies when coupled with a suitable camera and filters and used on a mounting driven to track the stars.

For visual astronomy, one usually needs a large-aperture telescope. Whether you want to see detail on the planets or find remote and faint galaxies, you need a wide aperture to bring in the light or to show fine detail. This is where reflecting telescopes excel, as their mirrors can be made in large sizes and give images free from the false colour of low-cost refractors. But there's one snag with mirror telescopes: their optical designs are difficult to make in wide-field versions with good definition across the field of view, other than at high cost. Photographs in particular may show stars as short streaks if they are more than a degree or so from the centre of the image.

Refracting optics, however, can be designed to give wide-field views with good definition from edge to edge in what photographers call 'fast' systems – that is, with comparatively short focal ratios such as f/5. The lower the f-number, the shorter the exposure time needed.

Many photographers with DSLR cameras already have a lens that will do this, such as a 70–200 mm zoom lens, and such lenses are available for a few hundred pounds either new or secondhand. While these are fine for everyday use, they usually have limitations when photographing stars, which are a severe test for a lens, at fast focal ratios. Images at the edges of the field of view may show distortion, and older lenses may show coloured haloes around bright stars. The usual solution is to stop the lens down using the built-in iris diaphragm, so that it operates at maybe f/8 instead of f/5. This improves the images somewhat, but at the cost of needing to give much longer exposure times.

Lens designs were transformed in the late twentieth century with the advent of new types of glass. It is possible to design lenses with virtually no false colour and generally improved characteristics by using more expensive types of glass in the design, usually referred to as ED (for extra-low dispersion) glass. A lens with ED in its description would be expected to use such glass, although this isn't a guarantee that there will be no false colour at all. Many photographic lenses now incorporate at least some ED or UD (ultra-low dispersion) glass.

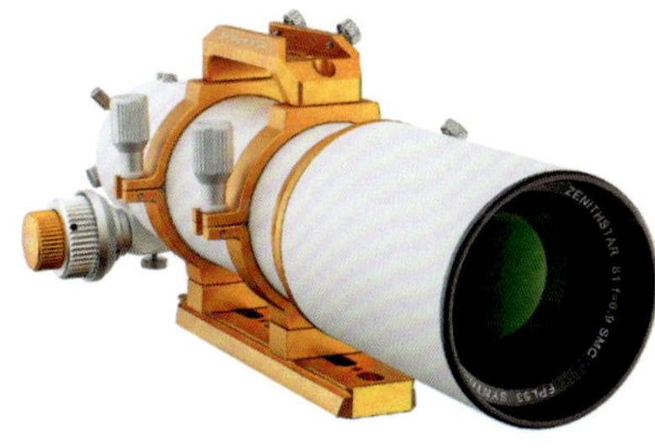

When used with a separately available focal reducer and field flattener, this William Optics Zenithstar 81 APO can be converted from an 81-mm aperture, 559-mm focal length, f/6.9 lens to a f/5.5 version. It comes with a transparent Bahtinov mask focusing aid that produces spikes on a star image that are symmetrical when the lens is in focus.

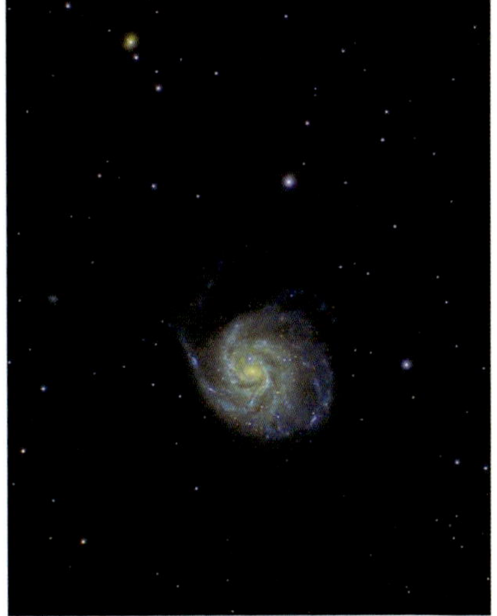

This image of galaxy M101 was taken using a Sky-Watcher 72ED 440-mm f/5.8 refractor. Lee Hore took 80 × 3-minute exposures over two nights from Cornwall to produce this result.

A photographic lens will have a standard fitting for use with your chosen make of DSLR camera, and will contain an iris diaphragm and probably many other features needed for everyday photography. However, refracting lenses for astronomy are usually designed to be used at full aperture only, and to be fitted to astro-cameras which have threaded nosepieces. A common size is referred to as M42, being 42 mm diameter, but bear in mind that there are two M42 threads – the original Pentax lens thread, with a pitch of 1 mm, and the T2 thread, with a

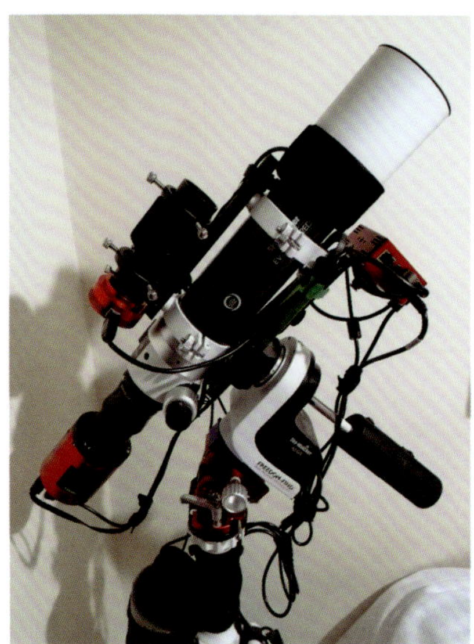

Lee Hore's set-up. The main camera is a ZWO ASI294MC colour astro-camera while the autoguider uses a ZWO ASI120MC camera. The red ASIAir box towards the top of the telescope tube allows exposures to be controlled from a smartphone. The mount is a Sky-Watcher Az GTi GoTo converted to equatorial mode.

pitch of 0.75 mm. If you mix the two up, you will damage both threads! Adapters are usually needed for DSLR cameras.

Another feature of the new breed of small refractors is that they are generally mounted on a standard dovetail for attachment to a wide range of telescope mountings. Many also have saddles for finder scopes or autoguiders – essential for much astrophotography. And, unlike photographic lenses, with suitable adapters they can be used with eyepieces for viewing through as telescopes, although don't expect to push them to magnifications over about 75.

The ads for small refractors usually contain jargon that might be confusing. Their lenses always have at least two separate components, called *elements*. A basic two-element lens is called an *achromat*, meaning that false colour is corrected to some extent. Many telescopes of all sizes use achromatic lenses, but unless at least one of the elements is made from ED glass or its equivalent they provide only a limited amount of colour correction and will give those coloured haloes around bright star images.

The next step up is an *apochromatic* lens, often referred to in lens descriptions as APO. These offer much better colour correction, although just because a lens is described as APO doesn't mean that it will give perfect images totally free from false colour or distortion across a wide field of view. Another term that you may encounter is *astrograph*, which implies that the instrument is designed specifically for astrophotography.

The other aspects of the set-up are a good mounting and a camera. While a simple tripod will support the instrument for short exposure times, for

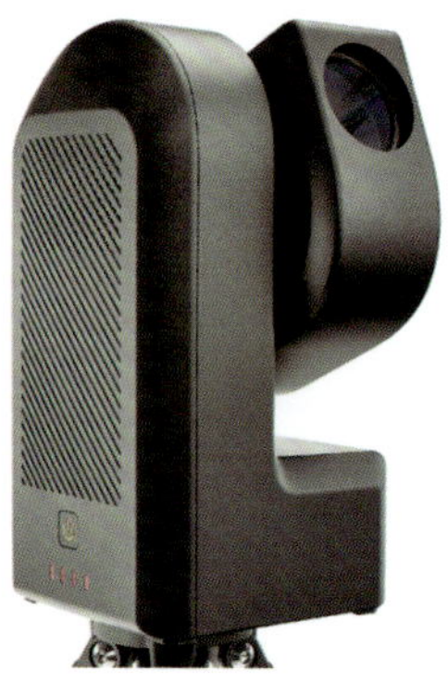

ZWO's Seestar 50, a 50-mm smart telescope, is creating a sensation among amateur astronomers for its ease of use in taking images and comparatively low cost. It comes with a solar filter and basic narrowband filter.

astro-imaging you need a mounting that will track the sky and allow exposure times of at least a minute or two. This could be a fairly simple tracking mount on a conventional tripod or a more advanced GoTo mounting, usually of the equatorial type, that will find any chosen object and track it. The longer the focal length of your telescope, the more precise the tracking needs to be. Fairly few telescope mounts will track objects in the sky precisely enough for long exposures, which is why autoguiders are needed. These are small telescopes with their own cameras, usually carried on top of your main telescope, and linked to the mounting. They automatically keep a star near your target centred on their own field of view, making corrections every few seconds if necessary.

As for the camera, you can attach a standard DSLR or mirrorless camera to the telescope, or you can choose a specific astro-camera. This usually requires a laptop to run the camera and view the images. However, astrophotographers are increasingly turning to minicomputer modules which attach to the telescope and allow you to control the whole astrophotography process, including finding the object, calibrating the autoguider and carrying out a series of exposures, over a Wi-Fi link to your smartphone or tablet. A market leader is the ZWO ASIAir, which works with ZWO, Canon and Nikon cameras.

Those in urban areas who despair of seeing stars let alone photographing deep-sky objects can still take amazing images by using narrowband filters in their system. The photos on pages 23, 41 and 77 were taken using these filters. They effectively cut out most of the light pollution, and even permit photography on nights of Full Moon!

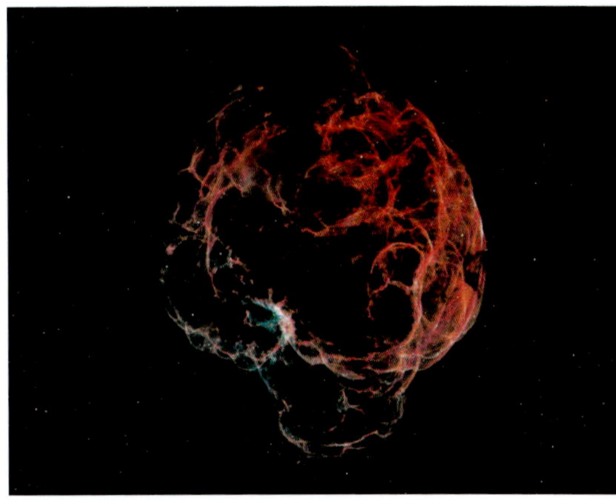

The faint supernova remnant Sh2-240, also known as the Spaghetti Nebula, spans some 3 degrees of sky between Taurus and Auriga, so only a wide-field instrument is able to get the whole object in a single frame. This image was taken from the heavily light-polluted London suburb of Ealing with an Askar FMA180 lens by John Davies. He used Antlia narrowband hydrogen alpha and oxygen III filters on a ZWO ASI2600MM Pro camera, with a total exposure time of 2 hours 15 minutes.

If all this seems too much bother, you can now buy the whole package of APO refractor, imaging system and GoTo mount in the form of a smart telescope, controlled directly from a smartphone or tablet. ZWO's Seestar 50 provides just this for around the same cost of a good compact camera. Put it down on level ground and within minutes it will have aligned itself and be ready to take images of any of the objects in its database. It has limitations compared with a custom-built set-up, of course, the most obvious being that you can't view through it with an eyepiece. But the message is that the small refractor is here to stay.

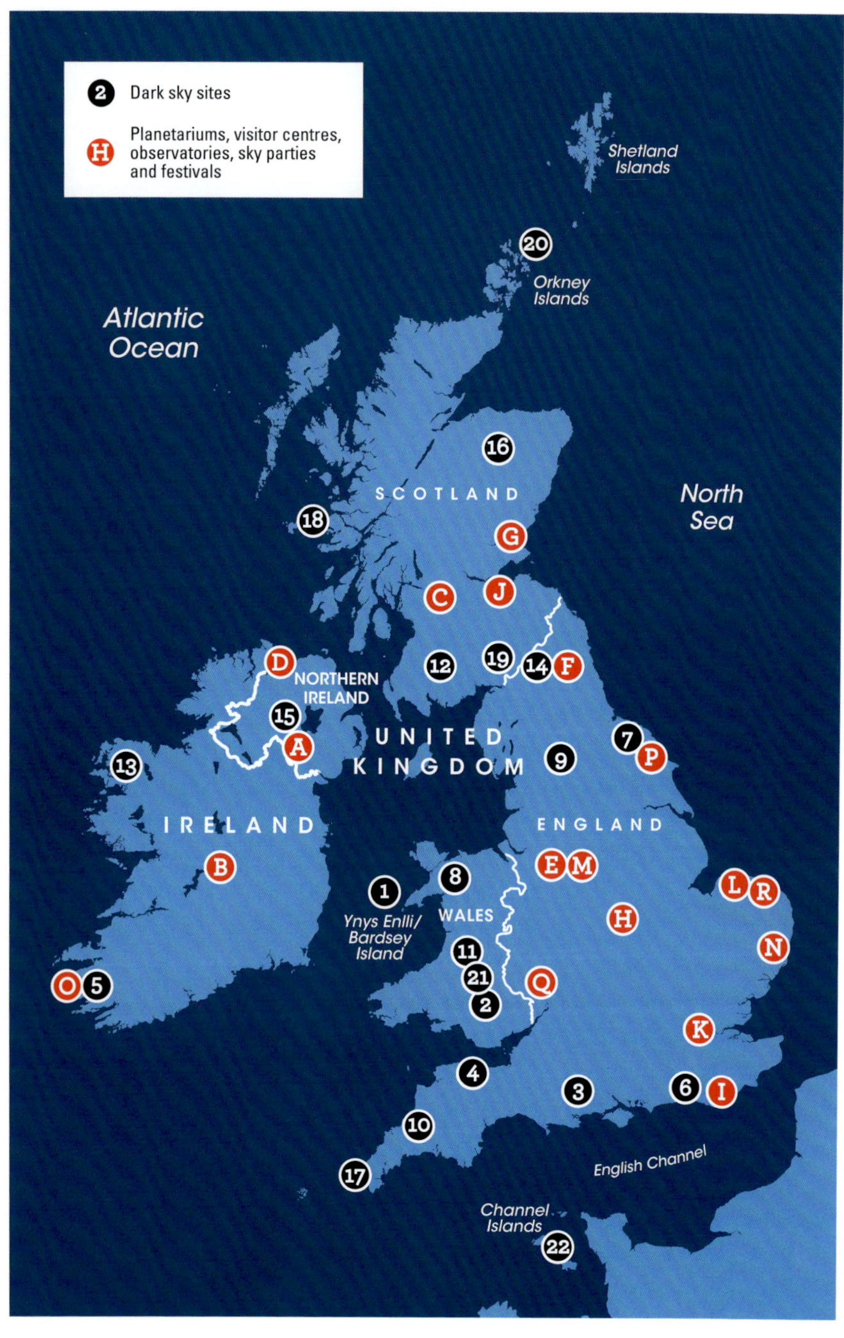

2 Dark sky sites
H Planetariums, visitor centres, observatories, sky parties and festivals
Shetland Islands
20
Orkney Islands
Atlantic Ocean
16
SCOTLAND
North Sea
18
G
C
J
D
12
19
14
F
NORTHERN IRELAND
15
A
UNITED KINGDOM
7
P
9
13
IRELAND
ENGLAND
B
E
M
1
8
L
R
Ynys Enlli/ Bardsey Island
WALES
H
11
N
O
5
21
Q
2
K
4
3
6
I
10
English Channel
17
Channel Islands
22

Take a break from your backyard stargazing! View the pristine heavens from a protected dark-sky site, visit a major observatory or planetarium, or join in the fun at a star party or astronomical festival.

DARK-SKY SITES

The International Dark-Sky Association, based in Tucson, Arizona, checks out the darkest places in the world. In these islands, 22 places are internationally recognised for their unsullied view of the heavens or their commitment to combating light pollution. Many are in National Landscapes, formerly Areas of Outstanding Natural Beauty.

DARK SKY SANCTUARY

Noted for exceptionally dark nights and a nocturnal environment protected for its scientific and cultural heritage, a Dark Sky Sanctuary is often in a remote location.

1. Ynys Enlli/Bardsey Island

Designated: 2023 • *Area:* 1.8 sq km
Nearest town: Pwllheli • *Website:* https://www.bardsey.org/darkskysanctuary
Europe's first designated Dark Sky Sanctuary is home to a major bird observatory, with an emphasis on nocturnal species. It lies 3 kilometres off the Welsh coast, and is accessible only by small boat during good weather.

DARK SKY RESERVES

Dark Sky Reserves are generally large, and consist of a very dark core region surrounded by a peripheral area where lighting is minimal. Eight out of a total of 19 worldwide are in Britain and Ireland:

2. Bannau Brycheiniog/Brecon Beacons National Park

Designated: 2013 • *Area:* 1347 sq km
Nearest towns: Brecon, Merthyr Tydfil
Website: https://www.breconbeaconsparksociety.org/
Situated in the mountains of South Wales, where sheep outnumber humans 30 to 1, this Reserve is home to 33,000 people – yet lighting is controlled so that the core zone has some of the darkest skies in Wales.

3. Cranborne Chase

Designated: 2019 • *Area:* 981 sq km
Nearest towns: Salisbury, Shaftesbury
Website: https://cranbornechase.org.uk/
Not far from Stonehenge, Cranborne Chase is a chalk plateau rising to 277 m at Win Green. This National Landscape has commanding views to the west and to some extent to the north.

4. Exmoor National Park

Designated: 2011 • *Area:* 693 sq km
Nearest towns: Barnstaple, Minehead, Taunton
Website: https://www.exmoor-nationalpark.gov.uk/exmoor-for-everyone/stargazing-and-dark-skies
The first Dark Sky Reserve site to be designated in these islands, Exmoor is a moorland area with much preserved history and many monuments within the 81-sq-km core zone.

5. Kerry

Designated: 2014 • *Area:* 700 sq km
Nearest towns: Kenmare, Waterville
Website: https://www.kerrydarkskytourism.com/
Its location between the Kerry Mountains and the Atlantic Ocean gives this Dark Sky Reserve natural protection from light pollution, yet it's readily accessible from the Wild Atlantic Way, the stunning tourist route that runs through the reserve.

6. Moore's Reserve (South Downs)

Designated: 2016 • *Area:* 1627 sq km
Nearest towns: Brighton, Portsmouth
Website: https://www.southdowns.gov.uk/

Named after the British astronomy populariser Sir Patrick Moore (1923–2012) who lived nearby, this reserve is sandwiched between London and the busy seaside resorts of Brighton and Worthing – yet it retains remarkably dark skies.

7. North York Moors National Park

Designated: 2020 • *Area:* 1440 sq km
Nearest towns: Scarborough, Whitby
Website: https://www.northyorkmoors.org.uk/
Despite its proximity to the busy tourist destinations of Whitby and Scarborough, the North York Moors is a largely deserted expanse of heather and bog moorland, with wide views of the night sky.

8. Eryri/Snowdonia National Park

Designated: 2015 • *Area:* 2132 sq km
Nearest towns: Harlech, Porthmadog
Website: https://www.snowdonia.gov.wales
The national park encompasses around 10 per cent of the total land area of Wales, and the darkest skies are to be seen from the rugged central area around Mount Snowdon (1085 m).

9. Yorkshire Dales National Park

Designated: 2020 • *Area:* 2180 sq km
Nearest towns: Hawes, Kirby Lonsdale
Website: https://www.yorkshiredales.org.uk/
The largest dark-sky site in Britain and Ireland, the Yorkshire Dales National Park boasts impressive waterfalls and caves, and brilliant night skies within reach of major cities like Leeds and Manchester.

DARK SKY PARKS

Smaller regions with exceptionally low light pollution are designated Dark Sky Parks. Currently numbering eight in Britain and Ireland, more are being added every year.

10. Bodmin Moor Dark Sky Landscape

Designated: 2017 • *Area:* 208 sq km
Nearest towns: Bodmin, Launceston, Liskeard
Website: https://cornwall-landscape.org/

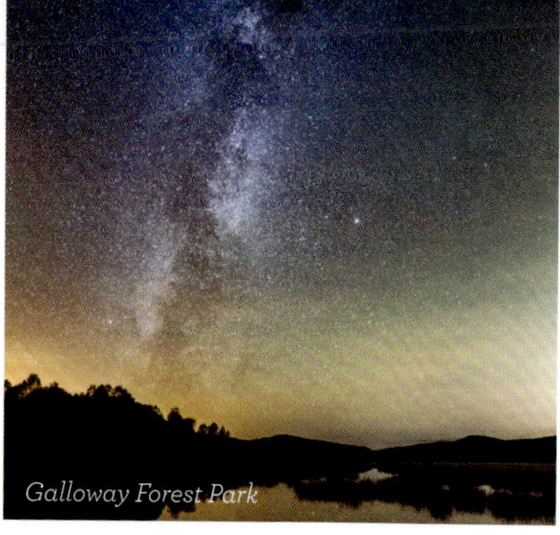

Galloway Forest Park

This remote, rugged area of granite moorland in north-east Cornwall is a working agricultural landscape, protecting it from large-scale development that would threaten dark skies.

11. Elan Valley Estate

Designated: 2015 • *Area:* 180 sq km
Nearest towns: Aberystwyth, Rhayader
Website: https://www.elanvalley.org.uk/
The city of Birmingham purchased this Welsh valley in 1892 to construct reservoirs that would provide a regular supply of pure water. Views of the starry night over the reservoirs are particularly impressive.

12. Galloway Forest Park

Designated: 2009 • *Area:* 780 sq km
Nearest towns: Girvan, Newton Stewart
Website: https://www.forestryandland.gov.scot/visit/forest-parks/galloway-forest-park/dark-skies
Galloway Forest Park is the largest forest park in the UK, and 20 per cent has been set aside as a core area with no illumination allowed. It's a mecca not just for astronomers but for nocturnal wildlife whose lives are often disrupted elsewhere by light pollution.

13. Mayo Dark Sky Park

Designated: 2016 • *Area:* 150 sq km
Nearest towns: Ballina, Castlebar
Website: https://www.mayodarkskypark.ie/
One of the largest expanses of peat landscape in Europe, this region supports a unique diversity of bog-dwelling species. Unsuitable for agriculture, and adjoining the Atlantic Ocean, the park should enjoy pristine skies far into the future.

14. Northumberland National Park and Kielder Water & Forest Park

Designated: 2013 • *Area:* 1592 sq km
Nearest towns: Jedburgh, Rothbury
Websites: https://www.northumberlandnationalpark.org.uk/; http://www.visitkielder.com/
Near Hadrian's Wall, built to keep the Picts from Roman Britain, this Dark Sky Park was designated as a bulwark against light pollution invading the darkness of northern England. It contains the largest reservoir and most extensive forest in northern Europe.

15. OM Dark Sky Park & Observatory

Designated: 2020 • *Area:* 15 sq km
Nearest towns: Cookstown, Magherafelt
Website: https://www.midulstercouncil.org/davaghforest
The first dark-sky place accredited in Northern Ireland, OM is set among rolling hills and is centred on the Bronze Age site of Beaghmore Stone Circles.

16. Tomintoul and Glenlivet, Cairngorms

Designated: 2018 • *Area:* 230 sq km
Nearest towns: Dufftown, Grantown-on-Spey
Website: https://www.cairngormsdarkskypark.org/
Containing the dramatic landscape of the Cairngorm Mountains, this Dark Sky Park is home to Britain's only herd of wild reindeer. And, if the weather is cloudy, the Park also contains the Glenlivet whisky distillery!

17. West Penwith

Designated: 2021 • *Area:* 136 sq km
Nearest towns: Penzance, St Ives
Website: https://cornwall-landscape.org
The very western tip of Cornwall, stretching down to Land's End, West Penwith is a wild landscape, with stunning sea views and ancient archaeological remains - some thought to be astronomically aligned.

DARK SKY COMMUNITIES

A town, village or complete island that's actively fighting light pollution can be designated a Dark Sky Community, with four listed in Britain and Ireland so far, plus one in the Channel Islands.

18. Coll

Designated: 2013 • *Area:* 77 sq km
Website: https://visitcoll.co.uk/dark_sky/
The Scottish island of Coll is home to just 200 permanent residents, plus myriad birds in its extensive nature reserve. The island has adopted a light-management plan to ensure its skies remain dark.

19. Moffat

Designated: 2016 • *Area:* 147 sq km
Website: https://visitmoffat.co.uk/
This former spa town is a tourist base for southern Scotland. It has strict outdoor lighting policies to reduce light pollution in its hinterland.

20. North Ronaldsay

Designated: 2021 • *Area:* 7 sq km
Website: https://www.northronaldsay.co.uk
The northernmost island in Orkney. Many visitors come to its bird observatory, and the island's appeal extends to astronomers with its Dark Sky Community designation.

21. Presteigne & Norton Dark Sky Community

Designated: 2023 • *Area:* 40 sq km
Website: https://darksky.org/places/presteigne-norton-dark-sky-community-wales
Two neighbouring towns in Wales combined forces to severely curtail night-time lighting in an area that's home to the Spaceguard Centre observatory and to colonies of light-sensitive bats.

22. Sark

Designated: 2011 • *Area:* 5 sq km
Website: http://www.sark.co.uk/
One of the Channel Islands, Sark was Europe's first Dark Sky Community. Its pitch-black skies are largely due to the island's ban on public lighting and motor vehicles apart from tractors.

PLANETARIUMS, VISITOR CENTRES AND OBSERVATORIES

For a great day out under the cloudless sky of a planetarium, a tour of a world-beating observatory or an evening observing the stars, here are some leading locations. Many venues combine a planetarium (), visitor centre (V) and observatory ().

A. Armagh Observatory and Planetarium 

Website: https://www.armagh.space/
Armagh has the longest-running planetarium and the second-oldest observatory in the UK. Combine a state-of-the-art digital planetarium show with a tour of the observatory's ancient instruments (book in advance for the latter).

B. Birr Castle Demesne 

Website: https://birrcastle.com/
Marvel at the well-preserved remains of the Leviathan of Parsonstown, the largest telescope in the world from 1845 to 1917. The Science Centre chronicles how the Third Earl of Rosse created his great instrument, and discovered the spiral shape of galaxies.

C. Glasgow Science Centre Planetarium 

Website: https://www.glasgowsciencecentre.org/discover/our-experiences/planetarium
One of Scotland's most popular visitor attractions, Glasgow Science Centre features a 15-m diameter planetarium. After viewing the night sky in comfort, you can be awed by the giant screen of the Centre's IMAX cinema.

D. Inishowen Planetarium 

Website: https://inishowenmaritime.com/planetarium/
Part of the Inishowen Maritime Museum, the planetarium is picturesquely set beside Lough Foyle. As well as putting on astronomy shows, the Maritime Museum uses the planetarium dome to project immersive ocean experiences.

E. Jodrell Bank Discovery Centre 

Website: https://www.jodrellbank.net/
Grab a close-up view of the Lovell Telescope, the great radio dish that tracked rockets in the early days of the space race, downloaded pictures from the Moon and now investigates the secrets of pulsars. Other displays include exhibitions, talks and interactive activities.

F. Kielder Observatory 

Website: https://kielderobservatory.org/
The Kielder Observatory, sited under some of the darkest skies in England, has a variety of telescopes for visual observing and astrophotography. It hosts some 700 events per year.

G. Mills Observatory 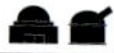

Website: http://www.leisureandculturedundee.com/culture/mills
On a hill above Dundee, the Mills Observatory was the UK's first purpose-built public observatory, and it carries on opening its doors to the public every clear weeknight. There's also a small planetarium and a gift shop.

H. National Space Centre 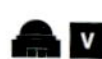

Website: https://spacecentre.co.uk/
Located on the outskirts of Leicester, this visitor centre focuses on the history – and future – of the UK in space exploration. Explore the wider Universe in the 192-seat Sir Patrick Moore planetarium.

I. Observatory Science Centre 

Website: https://www.the-observatory.org/
On a hillside at Herstmonceux in Sussex, the Observatory Science Centre is located within a cluster of green domes that once housed the telescopes of the Royal Greenwich Observatory. As well as fascinating exhibits, the observatory holds regular stargazing evenings.

J. Royal Observatory, Edinburgh 

Website: https://visit.roe.ac.uk/
Scotland's premier observatory no longer makes professional observations: its astronomers now build and use large telescopes on Hawai'i and in Chile. But the Edinburgh site is

still active, with a visitor centre, regular astronomical talks and public stargazing evenings.

K. Royal Observatory and Peter Harrison Planetarium 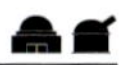

Websites: https://www.rmg.co.uk/royal-observatory; https://www.rmg.co.uk/whats-on/planetarium-shows

The home of British astronomy, the Royal Observatory at Greenwich is a fascinating museum of astronomy and timekeeping: stand on the Meridian Line, with a foot in each hemisphere! The planetarium hosts a variety of astronomical shows.

STAR PARTIES AND FESTIVALS

Enjoy an astronomical weekend or a festival with music – there's something for everyone at these annual events.

L. Autumn Equinox Sky Camp

Kelling Heath, Norfolk • September

Website: https://las-skycamp.org/

Claiming to be the largest star party in the UK, the Autumn Equinox Sky Camp fills three fields with astronomers, tents, trade stands and some of the top-end amateur telescopes.

M. Bluedot Festival

Jodrell Bank, Cheshire • July

Website: https://www.discoverthebluedot.com/

Astronomy's answer to Glastonbury, with live music, illuminated artworks, astronomy- and science-themed tents, cosmic inflated domes, family fun and talks from leading astronomers. Too busy and bright to actually observe the sky.

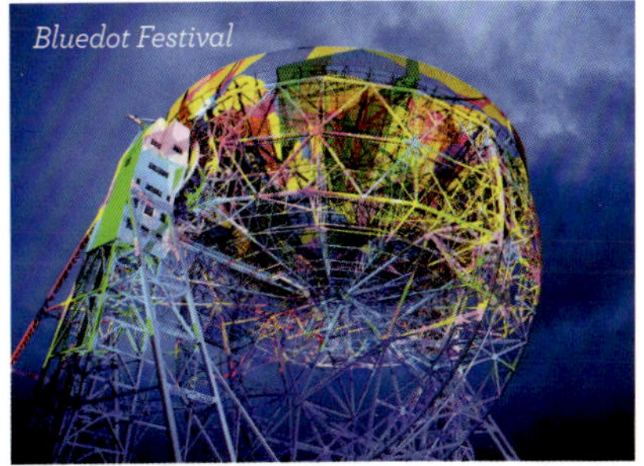

Bluedot Festival

N. Haw Wood

Saxmundham, Suffolk • April

Website: https://www.hawwoodfarm.co.uk/events/

Astronomers literally pitch up at Haw Wood farm for a small-scale, friendly week of stargazing, organised by local astronomical societies.

O. Skellig Star Party

Ballinskelligs, County Kerry • August

Website: https://skelligstarparty.com

Ireland's leading star party is held under the pristine skies of the Kerry Dark Sky Reserve. It features talks by leading astronomers as well as night-time observing.

P. StarFest

Dalby Forest, North Yorkshire • August

Website: http://www.scarborough-ryedale-as.org.uk/saras/starfest/about-starfest/

This event attracts astronomers from all over the country for a weekend of events that can include rocket-building, talks, an astronomical pub quiz and – of course! – observing the sky.

Q. Stargazers' Lounge Star Party

Lucksall Caravan and Camping Park, Herefordshire • October

Website: https://stargazerslounge.com/

Stargazers' Lounge is an online forum for amateur astronomers, but they meet in real life in Herefordshire for a weekend of talks, trade stands, socialising and skywatching.

R. WinterFest

Kelling Heath, Norfolk • December

Website: https://www.winterfestastro.co.uk/

Organised by the Birmingham Astronomical Society to give its members some respite from the city's lights, WinterFest is open to all who want to brave the December weather in pursuit of the glittering winter stars.

Many of the sites on pages 91–93 also host their own star parties or weekends: check their websites. Also see https://www.darkskiesnationalparks.org.uk/ for events.

THE AUTHOR

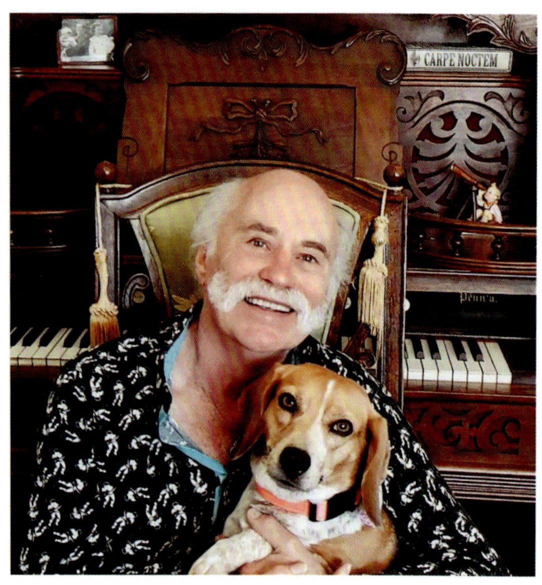

Nigel Henbest is an award-winning British science writer, specialising in astronomy and space. After research in radio astronomy at Cambridge, he became a consultant to both the Royal Greenwich Observatory and *New Scientist* magazine, and is a Fellow of the Royal Astronomical Society.

Originally with Heather Couper, Nigel has been writing Philip's *Stargazing* since 2005. The author of 50 other books, Nigel is a regular media commentator on breaking astronomy news. He co-founded a TV production company, where he produced acclaimed programmes and international series on astronomy and space.

Married with two stepdaughters, Nigel lives in Buckinghamshire and North Carolina. He is a Future Astronaut with Virgin Galactic, and asteroid 3795 is named 'Nigel' in his honour.

ACKNOWLEDGEMENTS

PHOTOGRAPHS

Front cover: Simon Hudson.
Jo Bourne: 47. **David A Cocklin:** 1. **Gary Cook/Alamy Stock Photo:** 92. **Commons Genuson (CC by-SA 3.0)/Wikimedia Commons:** 85a. **Davide De Martin & the ESA/ESO/NASA Photoshop FITS Liberator/NASA:** 85b. **John Davies:** 90. **Dave Eagle:** 59. **First Light Optics:** 86, 87b, 88b. **Shirlaine Forrest/WireImage/Getty Images:** 95. **Bob Franke/NASA:** 36. **Goddard/NASA:** 79. **Tracey Harty:** 71. **Bridget Henbest:** 96. **Nigel Henbest:** 7, 53. **Lee Hore, Delabole, Cornwall:** 87a, 88a. **Simon Hudson:** 77. **Peter Jenkins FRAS:** 41. **JPL-Caltech/NASA:** 31. **JPL-Caltech/NASA:** 67. **JPL-Caltech/SSC/NASA:** 83. **David Moug via Wikipedia:** 48. **Damian Peach:** 11. **Martin Ratcliffe:** 35. **T.A. Rector (University of Alaska Anchorage) and H. Schweiker (WIYN and NOIRLab/NSF/AURA):** 84. **Robin Scagell:** 6. **Ian Sproat/Mje Photography:** 17. **Franco Tognarini/Alamy Stock Photo:** 18. **UCLA/NASA:** 2. **Sara Wager, swagastro.com:** 23. **Denis Walsh:** 28. **Pete Williamson FRAS:** 64.

ARTWORKS

Star maps: Wil Tirion/Philip's with extra annotation by Philip's.
Planet event charts: Nigel Henbest/Stellarium (www.stellarium.org).
Pages 80–82: Chris Bell.
Page 90: Philip's